Science Communication and Public Engagement

Science Communication and Public Engagement

Evolving toward Science-Society Participation

Denisse Vásquez-Guevara

With
Judith McIntosh White,
David Weiss,
Angelica Ochoa-Aviles,
and Fernando Ortiz-Vizuete

LEXINGTON BOOKS
Lanham • Boulder • New York • London

Published by Lexington Books
An imprint of The Rowman & Littlefield Publishing Group, Inc.
4501 Forbes Boulevard, Suite 200, Lanham, Maryland 20706
www.rowman.com

86-90 Paul Street, London EC2A 4NE, United Kingdom

British Library Cataloguing in Publication Information Available

Library of Congress Cataloging-in-Publication Data Available

ISBN 9781666903430 (cloth) | ISBN 9781666903447 (epub)

To researchers, graduate students, and science advocates who believe in the power of sharing knowledge to evolve, heal, and transform reality. We thank our families and beloved ones who support our journey as faculty and researchers. Your love, patience, and understanding are the motors to keep us dreaming.

—Denisse, David, Judith, Angelica, and Fernando

Contents

List of Figures and Tables

FIGURES

TABLES

Acknowledgments

This book was made possible through the academic support that connects two nations and two universities, and five researchers who believe in the power of combining science and society. We acknowledge the kind support of the University of New Mexico in the United States and the University of Cuenca in Ecuador.

We want to acknowledge the unconditional support of the institutions that housed our past and current research projects and academic writing. In Ecuador, the Biosciences Department at the University of Cuenca, its current director Dr. Johana Ortiz, and the Nutrition, Health, and Physical Activity Research Group directed by Dr. Silvana Donoso, as well as the College of Philosophy, Letters, and Education for the initiative to support the creation of educational scholarship. In the United States, we want to thank the Communication and Journalism Department of The University of New Mexico, which unconditionally supports its faculty and alumni collaborations. Special thanks to Rowman & Littlefield and Lexington Books for offering us a space to create more knowledge about science communication for other researchers.

Special thanks also for the Ecuadorian Corporation for the Development of Research and Academia (CEDIA), and VLIR Network Ecuador for the space of developing multidisciplinary projects that apply science communication.

Introduction

Science Communication for a Complex World Scenario

Denisse Vásquez-Guevara

Science communication is still largely—and wrongfully—understood as a one-way dissemination strategy to inform audiences about scientific findings. However, a communication process is not complete if it does not flow in a two-way interchange to create mutual understanding and learning. Evidence shows that one-way science dissemination is frequently not effective for audiences (Metcalfe, 2020), in fact, it becomes problematic. Audiences frequently misunderstand data and its practical implications (Kohring, 2016; Schäfer, 2009). Additionally, people are disinterested in listening to or paying attention to science events and content that does not consider their needs, interests, and concerns, and they become reluctant to interact with researchers or be part of studies. However, we cannot blame them for this issue and their science literacy, or on scientists themselves. Instead, we want to invite our readers to reflect on how to change and improve audiences' interest and engagement with science.

Important efforts have been made to overcome unilateral science communication, by exploring in-depth the issues that emerge from traditional science dissemination strategies and their effects on audiences, such as their lack of interest and participation. These alternatives propose new paradigms and methods for research and practical initiatives based on participatory and collaborative methodologies (Bucchi & Trench, 2014). In this book, we depart from reflecting on how to use a participatory and dialogical approach to explore through multiple theories, methodologies, and case studies based on our work focused on how to shift science communication from mere dissemination to public engagement initiatives.

Still, science communication has a long way to go. Current and future researchers continue to have limited opportunities in their training programs that offer science communication courses; grant funding for science communication initiatives is very limited in North and Latin America, and it's barely starting to be considered by European grant agencies as a key component or through focused grant opportunities. Moreover, science communication is an expected initiative that researchers must do, but it is not recognized as part of their academic duties for hiring or promotion at universities or research institutions.

Science now, more than ever needs to be shared openly with people to provide evidence and data that can guide their everyday decision-making. Our daily life is full of decisions: parents that need to decide how to feed their families; women who do not know their options in case of violence; children who can learn from an early age that their actions have an impact on the environment; a patient who just got diagnosed with a chronic illness and does not know how to reorganize his/her life to adapt to treatments. For all these reasons, professors, and researchers need to explore how to invite people to dialogue about our research and acknowledge that individually we do not have all the answers, but together we can work for social change.

If you are a graduate student in a master's or doctoral program, or you are an experienced researcher we invite you to explore with us how to present your research findings to your local community, non-expert audiences, or your students.

OUR EXPERIENCE AS AUTHORS OF THIS BOOK

This book was written during a time of abrupt changes and challenges. The COVID-19 pandemic reshaped the life of humankind and suddenly rethinking how to develop new ways to keep systems functioning in an unknown and uncertain context became a priority. Society at large and people from diverse backgrounds, age groups, and education focused on survival through finding resilience strategies to face massive deaths, confinement, food access, and economic crises.

This reality was not foreign to the authors of this book. Like everyone in 2020, we were also in shock for several weeks. As university professors and researchers, we are spread across North and South America and have been working together for the last six years through research grants, outreach projects, and graduate programs. As the lockdown measures in nations worldwide impacted campus closures, all our academic lives migrated to our own homes. Our dining room table and home office became improvised research offices and classrooms. However, as academics, our role was not to observe and worry only but from our own knowledge reflect on how to navigate adversity, support our students, and propose solutions to address the crisis.

We recognize that daily life changed for everyone, and disparities deepened for the most vulnerable. While high-income and upper-middle-class people had access to the Internet and devices to develop a space for socializing, working, and learning, some middle- to low-income groups faced unemployment, inequality in access to food and health, and limited or no-education opportunities.

Now more than ever, science communication has become an even more necessary tool to make science useful for decision-making processes, and prevention, and to offer scientific data that informs society at large. In this context, as communication and science communication researchers our main goal in writing this book was to find effective strategies to adapt them in a resilient fashion to transform them into useful initiatives. Moreover, these alternative methods, theories, and cases can respond to the needs of scientists who want to implement science communication into their projects. In this book, we share several examples of our work and reflect on alternative initiatives of science communication for public engagement.

THE NEW NORMAL AND THE NEW SCIENCE COMMUNICATION SCENARIO

Amid these reflections and challenges, the context was not completely negative for science communication, although it was still highly complex. For example, the way mass media operated to inform society also changed. Before COVID-19, mass media offered almost no space for science. Sections for scientific content in television and print media were scarce and even inexistent. In 2020, this scenario changed, we could see that scientists' presence in the news increased. The media invited scientists to discuss ideas and guidelines to inform and educate audiences and provide guidelines informed by scientific evidence.

Not only mass media but also new media registered changes in how science communication was present in society. In 2020 only, social media registered over 2,680 posts about COVID-19 prevention addressing topics such as personal hygiene information and home quarantines (Nature Research Custom Media, 2021). In prevention campaigns, scientists' media presence became key to support the prevention of the contagion and spread of this lethal virus. The public opinion's attention turned to scientists and saw them as reliable sources to become informed. Moreover, scientists also initiated multiple efforts to deliver valuable research that contributes to supporting not only health solutions, but the derived issues that the pandemic left us, such as mental health issues, poverty and economic crises, famine, and racial and political tensions.

As science communication scholars we can also acknowledge that COVID-19 was a huge wake-up call to the role that science communication for public engagement has in social change. Science communication is a fairly new discipline, especially with the goals of public engagement, that has been evolving more dramatically in the last three decades. However, only during the last years of the 90s to the present did efforts to include people's voices and opinions in the process to communicate science begin to emerge.

HOW TO USE THIS BOOK

This book has five chapters. Chapter 1 explains how and why science communication issues exist and are reproduced frequently. To overcome these issues, our proposal is presented in chapter 2. Here we will offer ways to rethink the researcher's positionality in relation to collaborating with audiences and how self-reflexivity based on decolonial thought can be useful to build trust with audiences and self-critique productively our own beliefs that limit our relations with key audiences. In chapter 3, readers can explore how to navigate the persistent limitations of science communication in grant writing and through participatory methodologies. Specifically, you will learn more about how to include science communication initiatives in research grant writing. Later, chapter 3 dedicates a section to creating more understanding about the use of participatory action research for science communication projects, its genres, and tools to co-create, new initiatives or evaluate existing ones. Moreover, to offer you a compilation of hands-on experiences we will present several examples and case studies of our very own projects that were conducted through participatory collaborations. In chapters 2, 3, and 4, we will offer several case studies to reflect on how different theories and models can inform science communication research and practice. Moreover, we will reflect on several procedures, strategies, and ideas that offer examples of how science communication can develop practical initiatives that demonstrate the versatility and usefulness of the discipline.

Chapter 4 will explore two frequent scenarios of science communication: co-design and evaluation case studies. The co-design case study will present a study in which we created a new health education tool for young asthma patients. Instead, the evaluation will present a multiple-case study that evaluated two already existing health promotion programs that needed to be retooled to improve their audience engagement. Finally, chapter 5 will explore how to use audiovisual narratives for science communication in the age of social media.

This book brought together five authors, scholars, and faculty members with diverse experiences, and a common goal, to contribute to science communication scholarship, teaching, and application in real-life academic contexts.

In North America, from the United States, we have the contribution of Dr. Judith McIntosh White and Dr. David Weiss. McIntosh White has over fifteen years of experience teaching strategic communication. Her experience as a public information officer has given her research opportunities in strategic public communication of science, health, and agriculture, and entertainment education for public engagement in science. Her experience contributes to the analytical and methodological sections of this book where strategies and the methods to develop effective science communication initiatives. Dr. Weiss has vast experience in advertising for over two decades. Moreover, his research focuses on political communication, media theory, health communication, and strategic communication; the media and popular culture industries contribute to the theoretical foundations presented in this book to sustain a meaningful analysis that is well-informed by scholarship and models that shed some light on the most frequent issues of science communication.

From South America, in Ecuador, our authors are Dr. Denisse Vásquez-Guevara, Dr. Angelica Ochoa-Aviles, and MSC. Fernando Ortiz-Vizuete. Dr. Vásquez-Guevara has been part of ten multidisciplinary research projects in Latin America and the United States, where she has developed science communication initiatives. She has also trained scientists of different backgrounds in science communication. Through her experience, Dr. Vásquez-Guevara contributed to this book by coordinating authors and compiling the case studies, and examples. Moreover, she has developed the science communication guidelines for structuring messages, and strategies for effective public engagement. Dr. Ochoa-Aviles is a medical physician and dedicated her career to biomedical and public health research. She has directed multiple research projects and has experience in developing health promotion and health prevention initiatives in Latin America. Fernando Ortiz-Vizuete is the author of over ten academic articles and book collaborations. His educational background is in media journalism, cultural studies, and social sciences. His research focuses on audiovisual and transmedia narratives, social media, and public opinion.

We hope this book contributes to your scientific endeavors and supports researchers in approaching their communities and target populations to dialogue, interchange ideas, and find common interests to collaborate in egalitarian settings to build sustainable solutions.

BIBLIOGRAPHY

Bucchi, M., & Trench, B. (Eds.). (2014). *Handbook of public communication of science and technology*. Routledge.

Kohring, M. (2016). Misunderstanding trust in science: A critique of the traditional discourse on science communication. *Journal of Science Communication*. https://jcom.sissa.it/archive/15/05/JCOM_1505_2016_C00/JCOM_1505_2016_C04

Metcalfe, J. (2020). Chanting to the choir: the dialogical failure of antithetical climate change blogs. *Journal of Science Communication*, *19*(2), A04. https://doi.org/10.22323/2.19020204

Nature Research Custom Media. (2021). *Science communication in the COVID-19 pandemic*. https://www.nature.com/articles/d42473-020-00329-z

Schäfer, M. S. (2009). From public understanding to public engagement: An empirical assessment of changes in science coverage. *Science Communication*, *30*(4), 475–505. https://doi.org/10.1177/1075547008326943

Chapter 1

The Evolution of Science Communication

Denisse Vásquez-Guevara, David Weiss, and Judith McIntosh White

Science communication is a complex field, one whose origins, supporting theories, and conceptual debates must be explored if we are to understand its evolution and its prevailing paradigms. To do so, in this chapter, we summarize the history of science communication and, in particular, the communication theories that inform and provide criteria for its applications, which include choosing appropriate spokespeople to deliver science content to audiences; developing guidelines for the selection and design of scientific content that is comprehensible for non-expert audiences; and determining the most effective communication conduits through which to deliver science communication campaigns and individual messages.

In this chapter, we also explore alternative communication models for science communication and discuss a number of studies about the outcomes of the application of these models. Finally, we consider the discipline's current evolution, highlighting in particular relational dialectics theory as a base upon which to build twenty-first-century science communication approaches for successful public engagement.

THE HISTORY OF SCIENCE COMMUNICATION: EMERGENCE AND CONCEPTUAL EVOLUTION

The emergence and evolution of contemporary science communication reflects the work of scholars from all of the world's inhabited continents. Thus, rather than seeing science communication as the product of one particular place or one historical moment, scholars such as Massimiano Bucchi and his colleagues (Bucchi, 2008; Bucchi & Trench, 2014) argue that the evolution of science communication has been intimately related to the needs of societies and nations around the globe and at various times to foster development, innovation, and economic growth. These needs have been particularly acute during post-war crises and other crises, which have driven nations and their leaders to educate their citizens and motivate them to address their challenges through scientific and technological inventions and innovations. This section, then, will describe the different research streams and movements that have driven the emergence, evolution, and transformation of science communication.

The Historical Context of Science Communication: Research Movements and Practice

The roots of science communication can be traced to the nineteenth-century US movement then known as "science vulgarization" or "science popularization" (Lewenstein, 1992). What was at the time the work of multiple institutions such as scientific societies, science journalists, and government agencies dedicated to disseminating science to the public ultimately led to media devoted explicitly to covering as well as science sections in general media outlets such as newspapers and later in specialized science magazines (Whalen & Tobin, 1980).

In the 1920s and 1930s, considered the "golden age" of science popularization (Gascoigne & Schiele, 2020), science reporting emerged and bloomed thanks to the efforts of the Science Service,[1] which provided journalists with training in science writing. By 1934, a group of these trained journalists created the National Association of Science Writers (NASW), a group whose numbers had tripled by the beginning of World War II (Lewenstein, 1992).

The disruption of World War II slowed and almost halted the evolution of the field as scientists' freedom to communicate their findings to the American and European publics was strictly limited due to security concerns (Gascoigne & Schiele, 2020). After the war, US government leaders saw science communication as a strategic effort that was necessary to reorganize the world through science and technology, motivate scientific progress, and apply scientific findings for commercial and industrial purposes (Lewenstein, 1992).

During the second half of the twentieth century, science communication evolved through several successive phases of research and practice, phases described today as "scientific literacy"; "public understanding of science" (PUS); "from deficit to dialogue"'; and "public engagement with science and technology" (PEST), a phase also known as "upstream engagement" (Bowater & Yeoman, 2013; Stilgoe et al., 2014). These research movements and their practices were, inevitably, embedded in and thus products of their specific places, time, and historical events, as we explain in the remainder of this section.

Scientific Literacy. The scientific literacy movement focused on filling what Tichenor, Donohue, and Olien (1970) called the "knowledge gap" between scientists and the public, and to do so by imparting factual scientific information to the population. The term "science literacy" implies that a basic stock of scientific knowledge should be an essential component of public education. The four components of the scientific literacy movement were (a) promote knowledge about scientific facts via basic textbooks; (b) create an understanding of the scientific method; i.e., experimental change; (c) promote an appreciation of the positive outcomes of science and technology; and (d) reject superstitious beliefs (Gregory & Miller, 1998). Scientific literacy advocates focused on the evaluation of public knowledge of science, a practice that led to their labeling of the public as ignorant and describing this "ignorance" in terms of what they called "knowledge deficits." This led to what is known today as the "deficit model" science communication (Cortassa, 2016). In this transition from science communication developed through science literacy and the emergence of the deficit model a significant study was developed by Durant in the United Kingdom and the United States in 1989 and was published in *Nature* magazine for the first time (Durant, 1999). Durant's study was later reproduced in 1992 The results yielded comparative data. This study found that 66 percent of UK citizens had a relatively good knowledge of science in comparison to American citizens, who had only a 17 percent score in the same study that asked, through the Eurobarometer Quiz, about general knowledge of science and used true or false questions (Bowater & Yeoman, 2013). This study was used in the 1990s decade as a key argument to defend the validity of PUS as a model that promotes science literacy and educates the population.

Later, this study was critiqued because it focused only on the deficits and the lack of education in culturally different populations (Holliman et al., 2009) but did not provide solutions for adequately addressing or overcoming the problems that it identified.

Public Understanding of Science (PUS). The public understanding of science movement emerged in the mid-1980s, raising concerns about the public's attitudes toward science. The founding document of the PUS movement was *The Bodmer Report* (1985), authored by Sir Walter Bodmer, a member of the British organization The Royal Society.[2]

As did publications of the science literacy movement, *The Bodmer Report* recommended the improvement of science education in public schools in order to develop adequate understanding of science among the general public, and provided specific recommendations for doing so. However, unlike the work of the science literacy movement, *The Bodmer Report* thoroughly described the issues of the deficit model of science communication regarding the lack of learning opportunities provided to non-expert audiences. Consequently, the PUS movement's principal goal was to make scientific researchers' findings more understandable and accessible to members of the public. Nonetheless, the portrayal of people as "ignorant" about science persisted (Stilgoe & Wilsdon, 2014; Wilsdon et al., 2005).

After the publication of *The Bodmer Report,* several organizations—The Royal Society, The Royal British Association, and the UK Committee on the Public Understanding of Science (UK-COPUS)—collaborated on efforts to provide funding for speakers to address organizations (e.g., women's scientific associations were given presentations about the importance of preventive healthcare, based upon data from applied research projects) and initiated strategies for social change and health promotion.

Despite these advancements, the PUS movement faced multiple critiques. Several authors agreed that the movement reinforced two aspects: (a) that all knowledge and expertise belonged to scientists, and (b) that disseminating more scientific knowledge from scientists to the public would generate a more-significant appreciation of science in the broad society (Gregory & Miller, 1998). Consequently, science illiteracy was understood as a deficit of knowledge, and public lack of understanding of science was seen as a deficit of audiences, under the assumption that members of the public were ignorant (Bowater & Yeoman, 2013).

Science communication practices developed under the PUS paradigm built upon and perpetuated the deficit model by developing strategies that allowed only one-way communication processes. Initiatives developed by advocates of the PUS movement failed to encourage the relationship between science and its audiences in the general public (Stilgoe & Wilsdon, 2014). In response, beginning in the 1990s, scientists and governmental organizations in the UK, appreciating the need to develop new forms of science communication that would promote dialogue and interchange between science and society, turned away from PUS, declaring it a "backward-looking" approach to science communication (Mulder & Goedhart, 2009).

Several studies have shown that PUS reproduces the flaws of the deficit model. For example, in their article "Why Should We Promote Public Engagement with Science?" Stilgoe and Wilsdon (2014) questioned the PUS model and showed how the deficit approach continues to be used in initiatives developed in line with PUS guidelines. Specifically, they found that researchers continue to work from one-sided assumptions and argue that people do not know enough about science to understand scientific topics. Wynne (2014) similarly argued that the PUS framework approach has not encouraged a cordial relationship between science and society, because it portrayed people as ignorant and science as unchanging.

Nonetheless, some scholars today still see PUS as offering a framework for dialogic models in science communication as science first needs to be understood by audiences in order to be valued (Gregory & Miller, 1998; Lewenstein, 1992; Turney, 1996) (which will be discussed in more detail in the next section). In principle, the public understanding of science approach promotes dialogue and interchange between scientists and the public, with the goal of non-expert members of the broader public being able to learn and understand science. Nonetheless, scientists that advocate the PUS approach can be blamed for making untested assumptions about the knowledge levels of audiences rather than conducting studies or making other efforts to actually know their audiences better (Metcalfe, 2019).

The Transition from the Deficit Model to the Dialogue Model. As a response to PUS, the third movement, "deficit to dialogue," emerged as an effort to bring together science and society and reflect on how to transition from a deficit paradigm to one based on dialogue. This movement encouraged scientific researchers and practitioners of science communication to question PUS assumptions about science, the role of audiences, and the nature of understanding that comprised the deficit model.

This movement promoted dismantling assumptions about the ignorance of the public and the unchanging and universal nature of science. Social scientists shifted their focus from critiquing the deficit model and its outcomes to exploring the perceptions and needs of people concerning science. Their studies revealed that the image people had about science was misguided, mostly because of the application of the deficit model to the development of science communication initiatives (Stilgoe et al., 2014). Consequently, people were skeptical about science, and at the same time, their information needs concerning science content were not satisfied. At the same time, people wanted to hear more about science and be involved in the topics of their interest—and they wanted scientists to respond to their needs and interests. In summary, people wanted to make their voices heard by those in power and to join forces with scientists and their work in order to effect change in decision making and policymaking in a variety of areas (Wynne, 2014).

The dialogue movement and the emerging studies that led to it helped social scientists to understand that a better research approach to science communication would be one that was interdisciplinary. Supporting this claim were findings of research projects that intertwined complex issues that involved politics, science, and public uncertainty (Stilgoe et al., 2014; Wilsdon et al., 2005). Consequently, science communication research and practices would be required to develop a broader engagement with stakeholders in order to identify effective actions for achieving social change. In this process, productive dialogue among scientists, members of society, and policymakers around issues of interest was a primary goal—and the impetus for the development of the fourth movement: science communication for public engagement.

Science Communication for Public Engagement. This movement is centered on increasing public attention to and interest in science topics (Holliman et al., 2009). Its advocates have approached the research of science communication for public engagement from quantitative and qualitative standpoints; specifically, quantitative research tests the hypotheses of prior studies while qualitative research explores ways to plan and communicate about research projects across diverse disciplines (Holliman et al., 2009).

However, instead of primarily testing only theories, some studies argue that science communication must instead follow an inductive and exploratory qualitative research approach or a grounded theory approach (Vásquez-Guevara, 2019). This methodological suggestion is based on the work of by Barney Glaser and Anselm Strauss (1967), which argued that models of social life should be developed from the data of everyday experiences and should be integrated with existing literature and theory at various stages in the process (Hunger & Müller, 2016). Consequently, this movement, which is itself still evolving, has simultaneously led to the fifth movement, known as upstream engagement.

Upstream Engagement. This movement seeks to facilitate the discussion of innovation, science and technology, and society through science communication. To do so, it embraces several questions: Why technology/science? What needs does society have to know about a specific topic? How are these topics related to people's everyday lives? And perhaps most important, why is this topic necessary for people? In summary, upstream engagement demands

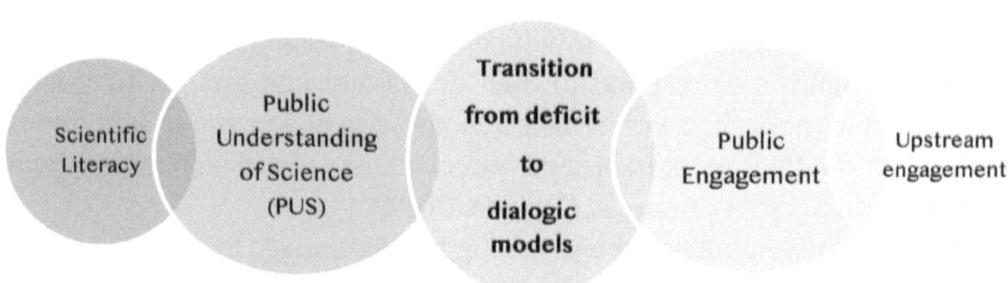

Figure 1.1. The chronological evolution of science communication. ***Developed by the authors.***

genuine openness and willingness on the part of scientists to not only dialogue with but also to truly engage with society in collaborative work through participation (Wilsdon et al., 2005).

This section explained how the evolution of science communication models the framework that guides their practices and research, usefulness, and critiques. A shared question among them remains unresolved: How can we bring scientists/researchers closer to society—and how can we overcome the scientific community's assumptions about audiences' knowledge levels and skills in order to help the public understand and engage with science content? In chapter 2, we will depart from a different starting point in order to explore how decolonial theories and methods can provide some guidelines to solve these issues and to develop more inclusive and engaging science communication initiatives.

COMMUNICATION THEORIES THAT INFORM SCIENCE COMMUNICATION

Science communication has transformed and evolved dramatically from its appearance to the current time (see figure 1.1). Contributing to its evolution and transformation are the various theories and models that have guided and shaped the discipline's research and practice. In this section, we will discuss some theories that have influenced and shaped science communication's goals and scope.

Science communication combines several communication theories into the models that guide its research and practice; as a result, its guidelines also vary. For this reason, it is important to discuss the reasons for the current gaps in the science communication literature and the concomitant variations in its applications. However, we do not see these variations in an exclusively negative light. Indeed, we suggest that combining different communication models and theories can guide scientists and communication professionals to plan and execute more effective and more engaging science communication initiatives for non-expert audiences.

Several authors from Europe, the United States, and Latin America have explored persistent challenges to successful science communication, such as misinformation, inaccuracy in science reporting, the lack of public participation in and engagement with science, among others (Bowater & Yeoman, 2013; Friedman et al., 1999; Hackett et al., 2008). These scholars agree that these issues are a consequence of the application of different theoretical frameworks that are the foundations of science communication models.

From Communication Theories to Science Communication Practice: Criteria for Choosing Spokespeople, Designing Messages, and Selecting Communication Conduits

Two-Step Flow Theory

The two-step flow theory was developed during the earliest days of television, in part in response to the concern that televised communication, and particularly political communication, had the potential to have a direct and outsized influence on the viewing audience. The two-step flow theory (Katz, 1957) and the empirical studies that supported its claims countered that concern by showing that politicians do not influence directly the majority of the citizens (or prospective voters) hearing their messages. Rather, political speech reaches and has the potential to persuade only those citizens who are particularly interested in it, and thus attending to it. These motivated listeners, categorized by Katz as "opinion leaders," in turn then pass along *political* messages (and their own opinions about them) to "opinion followers"—that is, their friends, co-workers, or family members who have lower degrees of interest in politics than they do. While two-step flow theory was developed to explain that *political* communication, specifically, thus goes through two stages in order to reach (most) citizens—first, the politician speaks (typically via media channels) to opinion leaders; second, opinion leaders communicate interpersonally with their acquaintances—it has since been shown to apply equally well to other topics around which people similarly divide themselves into opinion leaders and followers, such as sports, popular culture, and fashion, among others.

Two-step flow theory is especially useful for addressing the recurring problem of the accuracy of science content when conveyed by journalists or other individuals who act as opinion leaders in regard to scientific content (Dunwoody, 2014). When the science opinion leader is someone other than the original author/researcher, there is a higher risk that the opinion leader might shape—or misshape—the findings, possibly communicating inaccurate information to the public (Nisbet & Kotcher, 2009). But if scientists themselves are directly involved in the process of science communication to the public—that is, if they consciously and willingly take on the role of opinion leaders—then the problem of inaccurate science content communication might be solved. Katz himself demonstrated this in his famous "Elmira Study" (1957), in which he showed that a high level of expertise in a discipline raises the possibility of attracting and maintaining the attention of audiences.

Consequently, there are significant opportunities to create engagement and understanding through direct dialogue between researchers and their audiences. However, as Vásquez-Guevara (2019) showed, when applying the principles of two-step flow theory—a theory developed to explain communication via the media technologies available in the mid-twentieth-century United States—to twenty-first-century science communication practice in non-Western contexts, it is necessary to consider how the theory and its application need to be adapted and updated in order to successfully engage with members of the particular culture that scientists need to reach, using the media platforms—including social media networks—that are central to their lives today.

Framing

Although a communication theory, framing's foundations can be found in the fields of sociology and psychology (Borah, 2011). Frames act as interpretative schemas that allow audiences to make sense of an issue through images, words, phrases, and presentation styles. Put another way, frames can be understood as "persistent patterns of cognition, interpretation, and presentation of selection, emphasis, and exclusion" (Gitlin, 2003).

Frames are used by communicators to construct news or tell other sorts of stories about specific topics and, in so doing, to direct audiences to interpret them in certain ways (Druckman, 2001; Entman, 1993; Goffman, 1974). Framing theorists use the psychological concept of schemas of interpretation to explain how frames are used by communication professionals such as journalists, advertising copywriters, and public relations practitioners as devices that facilitate the organization of information and the packaging of events for audiences' interpretation.

Framing and Science Communication. Entman (1993) defined framing as the exercise of selecting aspects of a perceived reality and making them more salient in a communicative text. In science communication, then, framing can be used to guide researchers in constructing messages about their findings so as to make more salient those aspects that will maximize audience engagement (Listerman, 2010). Frames can define a problem and its causes and can provide moral evaluations and treatment recommendations (Gamson, 1992)—but must always be done while keeping in mind the specific goals of communicating the science content (Vásquez-Guevara, 2019).

As frames act as interpretative schemas that allow audiences to make sense of an issue (Entman, 1993), when designing science content, communication practitioners can identify relevant everyday situations and topics of interest to their target audience in order to create messages that are engaging to them. For example, in the case of science reporting, frames developed by journalists construct schemas to help audiences to put issues into shared and commonly understood contexts (White, 2013).

In the specific case of science communication research and practice, the available literature suggests that message framing not rely on anecdotal observation of audiences. Rather, for successful science content framing, formative research focused on target audiences is necessary (Bubela et al., 2009). The results of formative studies are crucial to identifying audience-appropriate topics, interests, and informational needs. Formative research, which will be explained in greater depth in chapter 3, can be developed that incorporates the use of qualitative methodologies such as focus groups or by using the guidelines of participatory action research (PAR).

Medium Theory

In the words of Joshua Meyrowitz, who, in the 1980s, coined the term and established its parameters, "medium theory is a special type of media study that focuses on [the] characteristics of each medium and on how each medium (or each type of media) is physically, socially, and psychologically distinct from other media" (2009, p. 518; see also Meyrowitz, 1985, p. 16). In focusing on the characteristics of each medium, medium theorists seek to understand how those characteristics "may have an influence on human interactions, institutions, and social structure" (Meyrowitz, 2009, pp. 518–519). These characteristics include the following:

- the type of sensory information the medium is able or unable to transmit;
- the form(s) of information conveyed by the medium in each sense;
- whether the medium offers unidirectional versus bidirectional versus multidirectional communication;
- the degree and type of control the users have over reception and transmission;
- the physical requirements for using the medium;
- the degree and type of human intervention/manipulation that is necessary or possible in creating a message;
- the scope and nature of dissemination of the medium (e.g., the question of how many people in different locations can experience the same message at the same time);
- the medium's durability . . . and portability; and
- the relative ease of difficulty of learning to code and decode messages in the medium. (Meyrowitz, 2009, p. 519)

Although the insights of medium theory and its close relative "media ecology" (Lum, 2006; Postman, 1985; Strate, 2006) have been available to communication scholars for nearly four decades, science communication scholars have only just begun in recent years to bear in mind the features and characteristics of various media conduits and how different media can be used to reach general audiences more effectively.

For example, climate change scientists have undertaken—and compared the results of—initiatives on platforms such as Twitter and Facebook to create dialogue with society (Pearce et al., 2015) and therefore consider the ways in which these two platforms' "biases" differ: to wit, Twitter is "biased" toward the communication of concise, primarily verbal messages whereas Facebook allows for longer, more in-depth, and frequently more multimodal communication—and blogs can provide spaces for even richer sources of data, interaction, and spontaneity for content creation. Bearing in mind the insights of medium theory also has opened the opportunity for scientists and the public to engage with new developments in science and policy and to develop interventions to foster discussion, contestation, and alternative networks (Pearce et al., 2015). Similarly, Collins and colleagues showed that the different structural characteristics of platforms such as Facebook, Instagram, Twitter, and YouTube suggest different ways to communicate science content, depending upon which social network is used by an intended audience (Collins et al., 2016).

More recently, Vásquez-Guevara (2019) conducted a case study in scientific communication for public engagement in which she explored the optimal ways to develop communication campaigns targeting teens and pre-teens about healthy eating and exercise practices. In developing the campaigns, she explicitly incorporated the insights of medium theory and thus focused not only on the content of their messages but also on the "biases" or characteristics of each potential medium that might deliver those messages.

Uses and Gratifications Theory

Rather than pointing the metaphorical camera at the medium, uses and gratifications theory focuses explicitly on the user or consumer of media and media content. Originally developed in the 1940s (Herzog, 1941) and then substantially expanded in the 1970s (Katz & Blumler, 1974; Katz, Blumler, et al., 1973; Katz, Haas, et al., 1973), uses and gratifications theory rejects claims that media have powerful or direct effects on unsuspecting or naïve audiences and instead "imagines audience members to be discriminating users of media," arguing that "individuals use media content actively to achieve their own goals rather than being passive consumers of it" (Littlejohn & Foss, 2011, p. 164).

Specifically, according to the theory (often abbreviated as "U&G"), audiences are active, goal-directed, and responsible for choosing media to meet their own needs and provide their own gratifications—although media are only one set of factors that audiences consult to achieve those gratifications (Littlejohn et al., 2011). Further, because audience members are individuals, each with different personalities, priorities, and needs, including media-provided needs—and because media must compete with each other for the attention of increasingly fragmented audiences—audience members "are assumed to have considerable agency with regard to media: they know their needs and how to gratify those needs" (Littlejohn et al., 2011, p. 164). This reality, then, drives message producers to create content that will be engaging to their intended audiences.

Although the key claims of uses and gratifications theory have been embraced by communication scholars in a wide variety of subdisciplines since the 1970s, a surprisingly small number of science communication studies have applied its insights. One such project—which actually incorporated (although did not acknowledge) aspects of medium theory alongside those of U&G—was conducted by Gelmez Burakgazi and Yildirim (2014), who investigated Turkish fourth- and fifth-graders' uses of mass media (TV, newspapers, Internet, magazines) and assessed the features of those media as sources for science learning. The authors found that students were "active in choosing and utilizing media to meet their cognitive, affective, personal integrative, and social integrative needs" (p. 168). They also reported that students' motivation to use particular media to achieve certain gratifications were influenced by messages' "accessibility, content, and presentation" (p. 168).

More recently, in the field of educational technology, Menon (2022) used U&G principles as the basis for an investigation of learners' motivations and intentions when using a variety of education apps (downloadable software programs integrated with learning materials, allowing students to access course material anywhere), an issue of increasing importance during the COVID-19 pandemic. Menon identified no fewer than seven gratifications sought by learners, with three of them—academic assistance, convenience, and social influence—identified as significant predictors of the intention to use such apps (2022).

Regarding science communication theory and models, several scholars argue that science communication models are not theories. As theory is defined as a generalized finding or conclusion to explain or analyze an event, phenomenon, or object. Additionally, theory can guide research through testing and building upon the findings of prior studies (Bucchi & Trench, 2014; Rubin, 2021). Consequently, in science communication it is more accurate to reflect on how models work through theory to develop (a) practical initiatives for interpreting reality or (b) reflecting on how to address different issues that remain unsolved through research.

Newer approaches to science communication suggest that models are useful frameworks for addressing complex relations among its research and practice, creating a communication initiative, or analyzing existing initiatives (Holliman et al., 2008; Leach et al., 2008; Pearce et al., 2015). Over the last decade, science communication scholars and practitioners have reflected on how to promote public engagement and how to build closer relationships among scientists and society.

NEW MODELS TO PROMOTE ENGAGEMENT FOR SCIENCE COMMUNICATION

As a response to the persistent issues of the deficit models, alternative models emerged from the paradigm of dialogue and participation among scientists and people. Several authors from Europe, the United States, and Latin America have explored the persistent issues of science communication, such as misinformation, inaccuracy in science reporting, the lack of participation of society, and the engagement of science and society, among others, that seem to be common and unsolved (Bowater & Yeoman, 2013; Bucchi, 2008; Friedman, Dunwoody, & Rogers, 1999).

Science communication scholars from Europe, the United States, and Latin America have explored the persistent issues of science communication, such as misinformation, inaccuracy in science reporting, and the lack of interest of society in the science (Bowater & Yeoman, 2013; Dunwoody, 2014; Friedman et al., 1999). Results of these studies evidence that the issues are a consequence of the application of different theoretical frameworks as foundations for science communication models. Therefore scholars agree that these issues are a consequence of the application of different theoretical frameworks that reinforce deficits in the science communication (Bucchi & Trench, 2014; Rubin, 2021).

Dialogic models were suggested as alternatives to overcome the issues of the deficit models. Different than deficit, dialogic models depart from the paradigm of interchange and egalitarian participation. Moreover, dialogue offers different perspectives on the complex relationship between the research and practice of science communication. Several valid considerations for dialogic models are that:

1. Communication takes place through a two-way conversation among scientists and the public, assuming that scientists are motivated to understand the needs, interests, and concerns of a particular audience and recognize that the local knowledge and experiences of audiences are valuable to address social problems (Metcalfe, 2019).

2. Dialogue-based models must focus on how to capture the attention of the public, decision-makers, and scientists to create initiatives that attract people to get involved to discuss topics and reach mutual understanding (Davies, 2019; Holliman et al., 2009).
3. Finally, dialogic models have the goal of developing tangible actions through scientific research data for engaging scientists and laypeople. Therefore members of the audiences who are directly affected by a particular issue become active actors who are attentive and active to work with researchers to create initiatives that address or work towards solving a particular issue of mutual interest (Wilsdon et al., 2005; Wynne, 2014).

Metcalfe (2019) in her doctoral dissertation makes an interesting approach that describes the tension among deficit and dialogue models as "blame models":

> . . . public may be being blamed for not understanding and appreciating science, and scientists may be blamed for not understanding the public, trying to have positive relationships with the public, or attempting to engage with them. In some cases, both scientists and the public are blamed for not engaging with each other and others to try and more deliberately solve important societal issues. In other words, instead of the models being helpful for explaining and informing science communication practice they can be used to criticize the efforts of practitioners. (p. 7)

Metcalfe's approach of blame models was developed for comparing empirically the science communication models against practice and initiatives outcomes, we can agree that models do serve practitioners and researchers as guides to develop different procedures and useful reflections according to the research topic, communication goals with audiences, and planning the strategies to approaching the public and to design engagement strategies.

Additionally, we propose that every model, even dialogue and participatory frameworks that aim to overcome the deficit models, has its own shortcomings regarding how to understand audiences, choosing appropriate engagement strategies, or proposing strategies for sustaining and motivating audiences' participation. There is no one model that fits all the needs of research initiatives, although, we also suggest, that future research in science communication might investigate and evaluate practical science communication initiatives that apply specific science communication models outcomes. We might suggest to science communication practitioners and researchers that combining models components that integrate information, participation, and interaction along with other theories, and epistemologies, that serve to gain more understanding about audiences and the complexity of the research projects topics themselves, can be a more effective approach to overcome overfocusing on deficit or dialogue models, but rather using their components to propose initiatives that contribute meaningfully to science communication for public engagement.

Relational Dialectics Theory: Alternatives to Rethink and Facilitate Public Engagement

To paraphrase and elucidate Marx, "the masses make history, but the elites write it." Such is surely the case with the science communication public sphere. Although Habermas focused on the evolution and decline of the eighteenth-century bourgeois public sphere (Habermas, 1991) public spheres—elite, bourgeois, and common—have flourished since the dawn of man. However, the history of such common public spheres has been neglected and even eradicated. Thus, in this discussion of alternative models for science communication, we will focus on the emergence in the twenty-first century of a new type of mediated public sphere, the YouTube science video.

As a public and academic scholar, our research has focused on the use of media to communicate scientific knowledge and to engage the public with the scientific process and its outcomes. In the past, Judith McIntosh White, our co-author, has dug into the role of newspapers in the coverage of science-based crises (2006) investigated the role of newspaper reporters and PIOs in engaging the public with topics such as vaccine hesitancy and legislative initiatives concerning health (White, 2012b; Weiss, White, Stohr, & Willis, 2015; White, 2014; White, Willis, & Stohr, 2013; White & Rutherford, 2012); sought to discover the best channels through which to influence adolescents' abusive alcohol behaviors (Morales et al., 2021; White, Morales, Bentley, & Weiss (under review); Shields et al., 2016); probed the potential interaction of museum exhibits with developing STEM-favorable attitudes in primary and secondary school children (White, 2018); and looked at the attitudinal differences between science reporters and PIOs (White & Wingenbach, 2013). Based on Dr. White´s experiences working for seven years as a public information officer covering science for a major research university, she even published a research-based study challenging Habermas himself as he claimed that PIOs are not journalists, but only focused public relations practitioners (White, 2012a).

But the greatest insights into how public information and real audience engagement work came through the production of two films and the public showings of them with the audiences they depicted and who played integral parts in the evolving scripts and video images are chosen and crafted in the final cuts of the films. True products of the participatory action research paradigm, the films told the story of (1) a rural New Mexico Hispano environmental activist (White & Garcia, 2012) and (2) of the gentrification of an Albuquerque Hispano neighborhood (White & Garcia, 2014). Working with these communities to tell their stories in their own words demonstrate that the future of successful science communication and engagement lies in liaising with and encouraging grassroots participatory public spheres around issues crucial to such communities' survival.

The lens of participatory action research has led Dr. White most recently to an investigation of the burgeoning YouTube universe of science videos. Although the YouTube channels presenting these videos are by no means homogeneous, most have commonalities. Most use a lecture format featuring one-way communication, and most are offered by either scientists/engineers working in industry or higher education or by secondary school teachers seeking to fill a gap created by traditional education formats and structures.

But what if this universe could be reconfigured to form the basis of interest-focused participatory public spheres? What if it could be expanded to include its audiences, letting them define what they need/want to see as content and having input into its presentation? That thought intrigues us, as we believe that such mediated science communication embedded within stakeholder public spheres could both expand and reconfigure the parameters of science communication and public engagement, moving beyond both conventional theories used to drive—unsuccessfully—efforts to bring more people into the realm of science knowledge and appreciation.

For such an effort to succeed, however, we believe that communication theories and methods themselves, as applied to the conundrum of efficient science communication/engagement must welcome new approaches, two of which are surely the application of Relational Dialects 2.0 (theory) and Entertainment Education (method) to the issue of facilitating real public spheres in this area.

Relation Dialectics 2.0

Relational Dialectics Theory identifies discursive tensions as the heart of meaning-making. This section will clarify RDT's contributions to individual efforts to engage public members with science and the development of a contemporary public sphere for science communication.

First posited as a theoretical understanding of interpersonal communication, RDT addresses contradictions underlying encounters between two individuals (Baxter & Montgomery, 1996). RDT's dialectic approach stresses that in dyadic conversations, multiple viewpoints intersect as participants articulate their own desires, opinions, and agenda. As conversations unfold, RDT highlights messier, illogical, and inconsistent unfolding relationship moments. The resulting analyses focus on the emergence of meanings from struggle between different, often opposing, discourses. Thus, meaning is produced through dialogue, including both literal conversations and contextual discourses actors are aware of through their socialization. The interpenetration of competing discourses constitutes social reality. Given that many attempts to communicate science occur as face-to-face or mediated encounters between scientist/science communicators and one or more listeners, RDT proves helpful in crafting more effective ways to manage such situations.

However, as part of the critical turn in RDT, scholar Mikhail Bakhtin developed dialogics, positing that social life comprises an open dialogue between many voices achieving simultaneous differentiation from—yet fusion with—each other, resulting in the construction of human experience through communication. Bakhtin and others modified RDT into RDT 2.0, focusing on the ideological saturation of language resulting in people's struggles with competing voices as they try to make meaning (Halliwell, 2015). Such a lens holds powerful implications for developing a contemporary public sphere for science communication, bringing into the arena a focus on power and making space for marginalized voices, as well as a fusion of public and private realms. This emergent public sphere holds promise in an age of recurring crises that demand society to more effectively address science.

Engagement and Participation of Audiences in Science Communication

As science communication needs alternative frameworks to explore how to engage broader audiences, also new theoretical and methodological paradigms are required for not repeating issues that perpetuate the science-society separation. But what does engagement mean? According to the Cambridge Dictionary, engage is "to interest someone in something and keep them thinking about it" and "to become involved or have contact, with someone or something" (Cambridge University Press, 2022). Borrowing these definitions to science communication, engaging audiences means attracting people's interest and gaining their attention with the goal to make them retain specific information from scientific research. Moreover, it means establishing an ongoing relationship among scientists and people by being involved around topics of mutual interest.

Therefore, engaging audiences around science implies staying on audiences' minds, and making them curious about getting involved in scientific topics as well as becoming part of science initiatives. For this purpose, in the current scenario of massive information received through multiple communication conduits by audiences, capturing people's interests is not an easy task.

In this chapter, we have learned so far that is impossible to assume that mass media has the same effects on people, either social media or digital media, events or activities can be assumed as the best communication conduits. Science communication is more variable and complex to address than choosing state-of-the-art conduits or messaging strategies; it implies reflecting on frameworks that can be adapted, combined, or tailored to the needs of each scientific topic, its goals, audiences, and contents. For this reason, this section will reflect on dialogical models, participatory communication, citizen science, and entertainment education models that provide broader room to reflect and propose creative initiatives that motivate and aim to enhance audience engagement in science communication.

Regarding the challenges for citizen science, we can agree, as former practitioners and researchers in urban studies, public health, and media studies, that sustaining people's interest and engagement along the processes of participatory research is a difficult task. In terms of funding, it is ethical and necessary to compensate participants properly for their contributions and time, although there are limited guidelines for participants'/co-participants' compensation. Few funding agencies in Latin America, North America, and Europe have opportunities for citizen science projects. As researchers, we had the luck and privilege to have these funding opportunities, although it is also sad and difficult that when projects finish their duration, non-expert participants who are committed and engaged in research initiatives and want to keep collaborating have fewer opportunities to do so over time, even as volunteers.

Citizen science needs sustained funding support that also provides communication tools and resources that ensure access to information, dialogue events, and engaging activities that sustain in the medium and long term the results of research and initiatives co-created among scientists and people.

PARTICIPATORY COMMUNICATION MODELS

Participatory Communication emerges from Paulo Freire's work in the 1970s. Freire, originally from Brazil, proposed a revolutionary concept in his book *The Pedagogy of the Oppressed,* where he initiates a critical analysis of the educational system and how hegemonic frameworks from Western theories create issues that reproduce inequality. One of these issues is the *banking education* (Freire, 2018), this Latin American educator uses this metaphor to explain how students—much as banking accounts— are compared to subjects that are depositaries of knowledge who are expected to memorize and accumulate information without questioning it or being able to apply it in different contexts (Varona Domínguez, 2020). To overcome banking education, Freire proposes *liberation pedagogy* as a pedagogic approach that explores the conceptual trilogy of wanting, knowing, and power through a process of empowering people of providing awareness to people to acknowledge that they have the capacity to build something they want to achieve through effort, collective learning and collaboration (Prieto Hernández, 2013).

Derived from the notions of liberation pedagogy, Freire develops the proposal of participatory communication. The paradigm of participatory communication departs from the notions of pedagogy with the goal of mutual learning among educators and their students through establishing an egalitarian and democratic communication setting for interchange and dialogue. Through participatory communication, Freire proposed a new communication epistemology that enabled educators to teach how to observe theories to explain them, and put them into practice through collective analysis and active participation of their audiences (Tufte et al., 2020).

Participatory Communication Principles

Participatory communication proposes the elements of access, participation, and self-management, a rubric developed by UNESCO in 1977 (J. Servaes & Malikhao, 2012):

- *Access.* This principle refers to the use of media for public service, based on offering the opportunity of available content to the public to choose from a variety of relevant programs and to empower the public with ways to provide feedback and to transmit their reactions or demands.
- *Participation.* This concept implies that the public is deeply involved in communicating and using different media conduits in order to participate. For example, audiences can get involved in a television show's planning and production processes or a communicational piece. However, participation can also happen through audience involvement by consulting in the decision-making process about content evaluation.
- *Self-management.* This principle implies the most advanced way of participation. Through self-management, the public has power in the decision-making process of media creation in the overall process. Moreover, it involves the co-design and co-creation of the public´s collaborative work in topic selection, message design, and overall content creation.

Besides education, participatory communication principles have been applied to initiatives of social change and had successful results in advocating for retooling health programs, policymaking processes, and creating initiatives that address and solve societal problems in The Americas such as health inequality, community organizing, and participatory projects that reunite scientists and non-experts (Fernández-Aballí Altamirano, 2020; Suzina & Tufte, 2020).

Participatory communication through dialogic tools allows people to reach collective solutions to their concerns and issues and apply them in practice (J. Servaes, 1996). For science communication, the participatory models offer an alternative framework that are worthy to explore for practical initiatives and formative research. As one of the principles of this model, the proposal of an egalitarian framework to reunite researchers and their audiences in dialogue and collaboration opens significant opportunities to retool the planning, creation and execution of engaging and useful science communication initiatives. Throughout this book, and in specific we will explore in chapters 2 and 3 how several studies carried by some of the authors of this book that applied the participatory communication model are examples that evidence how to overcome deficit issues.

Designing Science Communication in Research and Practice from a Participatory Approach

Participatory communication applied to science communication implies departing from the notion of active participation and collaboration, mutual benefits, and shared responsibilities among the parties. To create engagement, either in research or by designing practical initiatives, we need to reflect on the following guidelines.

- Cultural humility.
- Egalitarian participation.
- Co-designing initiatives.
- Execution or evaluation.
- Evaluation.

Citizen Science

Amid the search for developing alternative research frameworks, citizen science emerges as a paradigm that integrates scientists and non-experts in the design, planning, data collection process, and analysis of scientific research. As a term citizen science is a new concept. It first emerged in the 1990s decade, and it was addressed by Allan Irwin in his book *Citizen Science, A Study of People Expertise, and Sustainable Development*. Irwin's book proposed several reflections and guidelines to build participatory research projects that unite scientists and non-experts around collaborative inquiry. Among the main considerations, this book reflects on environmental studies to propose how scientists and people can collaborate ethically, promoting capacity-building strategies for non-experts to conduct data collection and analysis and communication considerations with diverse groups of non-expert audiences (Irwin, 2002).

As a concept, citizen science is understood as a process where the public participates along with researchers voluntarily in the scientific process, addressing real-world issues through collaborative research that may include developing research questions, conducting scientific experiments, collecting and analyzing data, interpreting results, making discoveries, developing technologies and applications, and solving complex problems (Holdren, 2015).

Almost two decades later, citizen science initiatives have multiplied across the world, from social sciences, public health, sociology, environmental studies. Moreover, science communication represents multiple opportunities for bringing collaboration between science and society in a bottom-up approach that reunites scientists and people around issues of common interest (Bonney, 2021). Today, citizen science has the power to motivate multigenerational audiences to not only engage but be active actors in issues that affect communities and society and learn new skills that can be useful for their lives in the case of adults of all ages. Particularly, we believe that citizen science has the power to engage young people who could be the next generation's faculty and researchers.

Nations across the world had also made important steps in policymaking for proposing laws that support citizen science. In 2017, The United States proposed the US Citizen Science and Crowdsourcing Act. European nations have also recognized the importance of citizen science and created the European Citizen Science Association, to create guidelines for conducting research and educating researchers, policymakers, and the public about procedures and guidelines for the inquiry (Hecker et al., 2018). Moreover, it is crucial to clarify that under the citizen science paradigm, the main goal is to go beyond publishing in academic journals or attending conferences and instead focus on building sustainable science-society collaborations that contribute to addressing with feasible initiatives real-world problems.

Citizen Science, Open Science, and Science Communication

As open science engages in advocating for accessible data, publications, and evaluation tools, citizen science has a role to play in open science and open access movements. As non-experts have limited access to literature, tools, and software, citizen science is driven through almost exclusively open science sources to develop training and research procedures (Hecker et al., 2018).

In citizen science initiatives, it is crucial to sustain multiple and interactive forms of science communication through the research process with mass media, funding agencies, and potential partners (Hecker et al., 2018). Moreover, it is essential to train researchers and citizen scientists or co-researchers in communication skills that can include new media, mass media, and interpersonal communication that promote dialogue and interaction to promote the research project's success and raise awareness among the target audiences and stakeholders. The current challenge for citizen science and science communication is to engage researchers as well as target audiences for getting actively involved in mutual learning and collaboration. Studies in the last two decades coincide that in all cases future science communication and citizen science initiatives need to combine multiway forms and conduits of communication that are innovative to attract scientists and target audiences (Constant & Roberts, 2017; Trench, 2006).

Ten Principles for Citizen Science

The European Citizen Science Association (ECSA) working group has developed ten principles to support the development of citizen science initiatives. These principles were the product of the collaborative work of museum representatives, universities, citizen science practitioners, and non-profit organizations that propose aims to guide how to develop responsible citizen science initiatives, these are the following guidelines (Hecker et al., 2018).

1. Citizen science projects actively involve citizens in scientific endeavour that generates new knowledge or understanding. Citizens may act as contributors, collaborators, or project leaders and have a meaningful role in the project.
2. Citizen science projects have a genuine science outcome. For example, answering a research question or informing conservation action, management decisions or environmental policy.
3. Both the professional scientists and the citizen scientists benefit from taking part. Benefits may include the publication of research outputs, learning opportunities, personal enjoyment, social benefits, satisfaction through contributing to scientific evidence, for example, to address local, national, and international issues, and through that, the potential to influence policy.
4. Citizen scientists may, if they wish, participate in multiple stages of the scientific process. This may include developing the research question, designing the method, gathering and analysing data, and communicating the results.
5. Citizen scientists receive feedback from the project. For example, how their data are being used and what the research, policy, or societal outcomes are.
6. Citizen science is considered a research approach like any other, with limitations and biases that should be considered and controlled for.

However, unlike traditional research approaches, citizen science provides opportunity for greater public engagement and democratisation of science.

7. Citizen science project data and metadata are made publicly available and where possible, results are published in an open-access format. Data sharing may occur during or after the project unless there are security or privacy concerns that prevent this.
8. Citizen scientists are acknowledged in project results and publications.
9. Citizen science programmes are evaluated for their scientific output, data quality, participant experience and wider societal or policy impact.
10. The leaders of citizen science projects take into consideration legal and ethical issues surrounding copyright, intellectual property, data sharing agreements, confidentiality, attribution and the environmental impact of any activities.

According to ECSA principles, citizen science core activities are engagement, involvement, collaboration and participation, equality, and transversal communication to all processes and stages in initiatives of research. Consequently, science communication becomes a crucial tool for successful projects in citizen science.

However, it is important to clarify that citizen science and science communication are not the same as or substitutes for each other, but rather convergent disciplines that intersect and support each other's goals and possible shortcomings. For this reason, we propose three criteria: scope and goals, communication, and accessibility, to better understand the differences and the convergencies between citizen science and science communication. In terms of scope, citizen science supports that communication must be integrated throughout the research process at all phases, instead of focusing on traditional frameworks, that are supported exclusively to communicate findings at the end of studies through one-way dissemination strategies (Wagenknecht et al., 2021). Moreover, in terms of communication, both citizen science and science communication must reflect on focusing it identifying clearly "which content is to be communicated, to whom, and how?" (Wagenknecht et al., 2021). Moreover, both disciplines seek to motivate interchange and mutual learning among scientists and audiences (Cooper, 2016). Although borders between science communication and citizen science can be blurry, science communication can be understood as external initiatives from the scientists, universities, or research institutes to the public, and all the internal scholarly communication that happens in the organization and coordination of research initiatives can be conceptualized as citizen science communication (Hecker et al., 2018). In terms of accessibility, science communication and citizen science must always consider how to propose an egalitarian framework for audiences' access and participation. Therefore, accessibility also means using tools, resources, and activities that are easy to access for audiences' use and participation (Bowater & Yeoman, 2013) (see table 1.1).

Entertainment Education Models

Entertainment education is more method than theory. Developed in Mexico by Miguel Sabado, it has been used extensively in the crafting of mediated public health interventions and, as such, can contribute to the emergence of a mediated public sphere based on shared entertainment experience.

Although early efforts to use entertainment-education methodologies to promote health focused on professional production of media to be transmitted over mainstream outlets, it has been increasingly adopted at the grassroots level since its introduction as a health promotion methodology. Entertainment education tools have been used to implement the ideas of Paolo Freire as applied to community health needs, with members of oppressed groups writing their own creative works to spread the word about indigenous wellness practices (Airhihenbuwa, 1995).

Numerous works by African health and culture researcher Collins Airhihenbuwa (1995) challenge Western paradigms that define health and wellness, disease, and proper healthcare, instead positing competing systems of the indigenous world, systems that can accommodate alternative public spheres in the areas of health promotion and community participatory action. Such systems notably incorporate music, dance, art, and ritual in healing efforts, focused on collective action and performativity, all aspects of both worship and entertainment. Similarly, Glanz, Rimer, and Viswanath (2008) identify the importance of contextualizing the evaluation of health behavior and taking into account audience settings and characteristics when planning health education and the crucial need to involve communities in this process.

Although the mechanisms of entertainment-education depend heavily on narrative (Petraglia, 2007), transportation, and such psychological phenomena as parasocial education (Brown & Basil, 2010; Papa et al., 2000), room undoubtedly exists to engage community members in crafting entertainment-education vehicles that serve their needs and promote health through the endemic cultural humility of group members. Thus, entertainment education methodology, often relegated to the mainstream Western area, could have much to offer in creating alternative mediated public spheres uniting members non-Western cultures in the search for appropriate health promotion.

Table 1.1. Citizen Science and Science Communication

	Differences	*Convergencies*
Scope and goals	Citizen Science refers to collaborative research that involves through an egalitarian framework, scientists, and audience experts in several research phases. Science communication informs, entertains, and educates the public about scientific content.	Both must develop engaging strategies to motivate the interest and involvement of scientists and audiences (Hecker et al., 2018).
Communication	Citizen science communication refers to all the internal communication processes to inform, coordinate, and dialogue among researchers and participants in citizen science initiatives. Science communication refers to all the external communication efforts from researchers to the society, target audiences that can use mass media and digital conduits.	The crucial questions that both disciplines need to reflect on before initiating planning are: "Which content is to be communicated? To whom? How?" (Wagenknecht et al., 2021) Dialogue is the common goal for both disciplines, fostering opportunities to interchange among scientists/researchers, research participants or collaborators, and broader target audiences (Cooper, 2016).
Accessibility	Citizen science must reflect on how to develop a sustainable dialogue among researchers and participants or co-researchers. Science communication must combine multiple communication conduits and strategies that promote interaction and mutual learning around scientific content.	Thinking about accessible communication conduits and activities for audiences' interaction and dialogue is crucial for both disciplines (Bowater & Yeoman, 2013). Accessible tools and resources for audiences' active involvement and collaboration in research and dialogue are also a crucial factor.

Developed by the authors.

THEORETICAL CONSIDERATIONS AND REMARKS FOR SCIENCE COMMUNICATION

In this chapter, we have developed a theoretical discussion to reflect on the persisting conflicts and issues in science communication. As authors, we believe that understanding the causes of these issues will provide meaningful knowledge that can guide future research and practical initiatives.

In order to develop innovative studies and initiatives, we also introduce the concepts of participatory action research methodologies and newer theoretical models, such as citizen science and entertainment education models, which could be useful for overcoming limitations in audience participation in scientific research. Moreover, these models have demonstrated its effectiveness in prior studies and initiatives in public outreach and engagement.

In chapter 2, complementarily we will deepen the discussion on how democratizing and ensuring audiences' participation in science communication initiatives.

NOTES

1. The Science Service was established in 1920. This organization was a news service designed to popularize science and disseminate scientific knowledge to the public. The Science Service was founded through the efforts of E.W. Scripps in collaboration with the American Association for the Advancement of Science (AAAS), the National Academy of Sciences (NAS), and the National Research Council (NRC) (Smithsonian, 2019).

2. The Royal Society is the oldest scientific organization in the world. It was founded in 1660 by Robert Boyle, Christopher Wren, and William Petty in the United Kingdom. The Royal Society led revolutionary advancements for scientific research and science communication through the publication and dissemination of science for non-experts (The Royal Society, 2022).

BIBLIOGRAPHY

Airhihenbuwa, C. O. (1995). *Health and culture: Beyond the Western paradigm*. Sage.

Baxter, L., & Montgomery, B. (1996). *Relating: Dialogues and dialectics.* Guilford Publications.

Bonney, R. (2021). Expanding the impact of citizen science. *BioScience*, *71*(5), 448–451.

Borah, P. (2011). Conceptual issues in framing theory: A systematic examination of a decade's literature. *The Journal of Communication*, *61*(2), 246–263. https://doi.org/10.1111/j.1460–2466.2011.01539.x

Bowater, L., & Yeoman, K. (2013). *Science communication*. Wiley-Blackwell.

Brown, W. J., & Basil, M. D. (2010). Parasocial interaction and identification: Social change processes for effective health interventions. *Health Communication, 25,* 601–602.

Bubela, T., Nisbet, M. C., Borchelt, R., Brunger, F., Caufield, T., Critchley, C., Einsiedel, E., Geller, G., Gupta, A., Hampel, J., Hyde-Lay, R., Jandciu, E. W., Jones, S. A., Kolopack, P., Lane, S., Lougheed, T., Nerlich, B., Ogbogu, U., O'Riordan, K., Ouellette, C., . . . Willemse, L. (2009). Science communication reconsidered. *Nature Biotechnology*, *27*(6), 514–518. https://doi.org/10.1038/nbt0609–514

Bucchi, M. (2008). Of deficits, deviations, and dialogues: Theories of public communication of science. In M. Bucchi & B. Trench (Eds.), *Handbook of public communication of science and technology* (pp. 71–90). Routledge.

Bucchi, M., & Neresini, F. (2008). Science and public participation. In E. J. Hackett, D. Amsterdamska, J. Wajcman, & M. Lynch (Eds.), *The handbook of science and technology studies* (pp. 449–472). The MIT Press.

Bucchi, M., & Trench, B. (Eds.). (2014). *Handbook of public communication of science and technology*. Routledge.

Cambridge Dictionary. (2022). Engage Meaning. Cambridge University Press. https://dictionary.cambridge.org/dictionary/english-spanish/engage

Collins, K., Shiffman, D., & Rock, J. (2016). How are scientists using social media in the workplace? *PloS One*, *11*(10), e0162680. https://doi.org/10.1371/journal.pone.0162680

Constant, N., & Roberts, L. (2017). Narratives as a mode of research evaluation in citizen science: understanding broader science communication impacts. *Journal of science communication*, *16*(4), A03.

Cooper, C. (2016). *Citizen science: How ordinary people are changing the face of discovery*. Abrams.

Cortassa, C. (2016). In science communication, why does the idea of a public deficit always return? The eternal recurrence of the public deficit. *Public Understanding of Science,* 25(4), 447–459.

Davies, S. R. (2019). Science communication as emotion work: Negotiating curiosity and wonder at a science festival. *Science as Culture*, *28*(4), 538–561. https://doi.org/10.1080/09505431.2019.1597035

Druckman, J. N. (2001). Evaluating framing effects. *Journal Of Economic Psychology*, *22*(1), 91–101. https://doi.org/10.1016/s0167-4870(00)00032–5

Dunwoody, S. (2014). Science journalism: Prospects in the digital age. In M. Bucchi & B. Trench (Eds.), *Handbook of public communication of science and technology* (pp. 43–55). Routledge.

Durant, J. (1999). Participatory technology assessment and the democratic model of the public understanding of science. *Science and Public Policy, 26*(5), 313–319.

Entman, R. M. (1993). Framing: Toward clarification of a fractured paradigm. *The Journal of Communication*, *43*(4), 51–58. https://doi.org/10.1111/j.1460–2466.1993.tb01304.x

Fernández-Aballí Altamirano, A. (2020). The importance of Paulo Freire to communication for development and social change. In Jan Servaes (Ed.), *Handbook of communication for development and social change* (pp. 309–327). Springer Singapore. https://doi.org/10.1007/978-981-15-2014-3_76

Freire, P. (2018). The banking concept of education. In E. B. Hilty (Ed.), *Thinking About Schools* (pp. 117–127). Routledge. https://doi.org/10.4324/9780429495670–11

Friedman, S. M., Dunwoody, S., & Rogers, C. L. (1999). *Communicating uncertainty: Media coverage of new and controversial science*. Routledge.

Gamson, W. (1992). Media images and the social construction of reality. *Annual Review of Sociology*, *18*(1), 373–393. https://doi.org/10.1146/annurev.soc.18.1.373

Gascoigne, T., & Schiele, B. (2020). Introduction: A global trend, an emerging field, a multiplicity of understandings: Science communication in 39 countries. In *Communicating science: A global perspective* (pp. 1–14). ANU Press. https://doi.org/10.22459/cs.2020.01

Gelmez Burakgazi, S., & Yildirim, A. (2014). Accessing science through media: Uses and gratifications among fourth and fifth graders for science learning. *Science Communication, 36*(2), 168–193.

Gitlin, T. (2003). *The whole world is watching: Mass media in the making and unmaking of the New Left*. University of California Press. https://play.google.com/store/books/details?id=00iwHPO73mkC

Glanz, K., Rimer, B. F., & Viswanath, K. (2008). *Health behavior and health practice: Theory, research, and practice.* Jossey Bass.

Glaser, B. G., & Strauss, A. L. (1967). *The discovery of grounded theory: Strategies for qualitative research*. Aldine.

Goffman, E. (1974). *Frame analysis: An essay on the organization of experience*. Harvard University Press. https://psycnet.apa.org/fulltext/1975-09476-000.pdf

Gregory, J., & Miller, S. (1998). *Science in public*. Perseus Books.

Habermas, J. (1991). *The structural transformation of the public sphere: An inquiry Into a category of bourgeois society.* MIT Press.

Hackett, E. J., Amsterdamska, O., Lynch, M., & Wajcman, J. (Eds.). (2008). *The handbook of science and technology studies*. MIT Press.

Halliwell, D. (2015). Extending relational dialectics theory: Exploring new avenues of research. *Annals of the International Communication Association, 39,* 1.

Hecker, S., Haklay, M., Bowser, A., Makuch, Z., Vogel, J., & Bonn, A. (2018). *Citizen science: Innovation in open science, society and policy.* UCL Press. https://doi.org/10.14324/111.9781787352339

Holdren, J. P. (2015). Addressing societal and scientific challenges through citizen science and crowdsourcing. Memorandum to the Heads of the Executive Departments and Agencies, Office of Science and Technology Policy. Executive Office of the President.

Holliman, R., Collins, T., Jensen, E., & Taylor, P. (2009). *ISOTOPE: Informing science outreach and public engagement. Final report of the NESTA-funded ISOTOPE Project*. http://oro.open.ac.uk/20090/

Holliman, R., Whitelegg, E., Scanlon, E., Smidt, S., & Thomas, J. (2008). Models of science communication. In R. Holliman, E. Whitelegg, E. Scanlon, S. Smidt,

& J. Thomas (Eds.), *Investigating science communication in the Information Age: Implications for public engagement and popular media* (p. 288). Oxford University Press; Open University Press. http://shura.shu.ac.uk/id/eprint/2816

Holliman, R., Whitelegg, E., Scanlon, E., Smidt, S., & Thomas, J. (Eds.). (2009). Practising science communication in the Information Age: Theorising professional practices (Vol. 2). Oxford University Press. *Science Communication, 31*(1), 127–129. https://doi.org/10.1177/1075547009339050

Hunger, I., & Müller, J. (2016). Barney G. Glaser/Anselm L. Strauss: The discovery of grounded theory. Strategies for qualitative research, Aldine publishing company: Chicago 1967, 271 S. (dt. Grounded theory. Strategien qualitativer forschung, Bern: Huber 1998, 270 S.). In S. Salzborn (Ed.), *Klassiker der Sozialwissenschaften* (pp. 259–262). Springer Fachmedien Wiesbaden. https://doi.org/10.1007/978-3-658-13213-2_59

Irwin, H. (2002). Is scientific investigation of postmortem survival an anachronism?: The demise of the survival hypothesis. *Australian Journal of Parapsychology*, *2*(1), 19–27.

Katz, E. (1957). The two-step flow of communication: An up-to-date report on an hypothesis. *Public Opinion Quarterly*, *21*(1), 61. https://doi.org/10.1086/266687

Katz, E., Blumler, J. G., & Gurevitch, M. (1973). Uses and gratifications research. *Public Opinion Quarterly*, *37*(4), 509–523.

Katz, E., Blumler, J. G., & Gurevitch, M. (1974). *The uses and gratifications approach to mass communication*. Sage.

Katz, E., Haas, H., & Gurevitch, M. (1973). On the use of the mass media for important things. *American Sociological Review*, 164–181.

Leach, J., Yates, S., & Scanlon, E. (2008). *Models of science communication* (pp. 128–146). Open University Press.

Lewenstein, B.V. (1992). The meaning of public understanding of science in the United States after World War II. *Public Understanding of Science*, *1*(1), 45. https://iopscience.iop.org/article/10.1088/0963-6625/1/1/009/meta

Listerman, T. (2010). Framing of science issues in opinion-leading news: International comparison of biotechnology issue coverage. *Public Understanding of Science (Bristol, England)*, *19*(1), 5–15. https://doi.org/10.1177/0963662505089539

Littlejohn, S. W., & Foss, K. A. (2011). *Theories of human communication* (10th ed.). Waveland Press, Inc.

Lum, C. M. K. (Ed.). (2006). *Perspectives on culture, technology, and communication: The media ecology tradition*. Hampton Press.

Menon, D. (2022). Uses and gratifications of photo sharing on Instagram. *International Journal of Human-Computer Studies, 168*, 102917.

Metcalfe, J. (2019). Comparing science communication theory with practice: An assessment and critique using Australian data. *Public Understanding of Science (Bristol, England)*, *28*(4), 382–400. https://doi.org/10.1177/0963662518821022

Meyrowitz, J. (1985). *No sense of place: The impact of electronic media on social behavior*. Oxford University Press.

Meyrowitz, J. (2009). We liked to watch: Television as progenitor of the surveillance society. *The Annals of the American Academy of Political and Social Science*, *625*(1), 32–48. https://doi.org/10.1177/0002716209339576

Morales, G. I., White, J. M., German, L., Dozal, M., Bentley, J., & Weiss, D. (2021). Binge-drinking prevention campaigns targeting Hispanic/Latino college students: Importance of individual attitudes and peer support. *Cogent Social Sciences, 7*(1), 1929681, DOI: 10.1080/23311886.2021.1929681

Mulder, H. A. J., & Goedhart, M. J. (2009). Book Review: Holliman, R., Whitelegg, E., Scanlon, E., Smidt, S., & Thomas, J. (Eds.). (2009). Investigating science communication in the Information Age: Implications for public engagement and popular media (Vol. 1). Oxford: Oxford University Press. Holliman, R., Whitelegg, E., Scanlon, E., Smidt, S., & Thomas, J. (Eds.). (2009). Practising science communication in the Information Age: Theorising professional practices (Vol. 2). Oxford: Oxford University Press. *Science Communication*, *31*(1), 127–129.

Nisbet, M. C., & Kotcher, J. E. (2009). A two-step flow of influence? *Science Communication*, *30*(3), 328–354. https://doi.org/10.1177/1075547008328797

Papa, M. J., Singhal, A., Law, S., Pant, S., Sood, S., Rogers, E. M., & Shefner-Rogers, C. L. (2000). Entertainment-education and social change: An analysis of parasocial interaction, social learning, collective efficacy, and paradoxical communication. *Journal of Communication, 50,* 31–55.

Pearce, W., Brown, B., Nerlich, B., & Koteyko, N. (2015). Communicating climate change: conduits, content, and consensus. *Wiley Interdisciplinary Reviews. Climate Change*, *6*(6), 613–626. https://doi.org/10.1002/wcc.366

Petraglia, J. (2007). Narrative Intervention in behavior and public health. *Journal of Health, 12*(5), 493–505.

Postman, N. *Amusing ourselves to death*. Penguin.

Prieto Hernández, A. M. (2013). Pedagogía y prácticas emancipadoras: Actualidades de Paulo Freire. *Innovación Educativa (México, DF)*, *13*(63), 155–160. http://www.scielo.org.mx/scielo.php?pid=S1665-26732013000300010&script=sci_arttext

Rubin, A. (2021). A look at the complex issues of science communication: The Routledge handbook of public communication of science and technology. *Journal of Science Communication*, *20*(5), R02. https://doi.org/10.22323/2.20050702

Servaes, J. (1996). Participatory communication (research) from a Freirian perspective. *Africa Media Review*, *10*, 73–91.

Servaes, J., & Malikhao, P. (2012). Advocacy communication for peacebuilding. *Development in Practice*, *22*(2), 229–243.

Shields, J. D., Archiopoli, A. M., Bentley, J. M., Weiss, D., Hoffmann, J., White, J. M., Sharp, M. K., Hong, Z., & Kimura, M. (2016). Binge-drinking attitudes and behaviors among Hispanic and non-Hispanic college students: Suggestion for tailoring health campaign messages. *Hispanic Journal of Behavioral Sciences, 38*(2), 243–263.

Smithsonian. (2019). *Science Service*. Smithsonian. https://www.si.edu/spotlight/science-service

Stilgoe, J., Lock, S. J., & Wilsdon, J. (2014). Why should we promote public engagement with science? *Public Understanding of Science (Bristol, England)*, *23*(1), 4–15. https://doi.org/10.1177/0963662513518154

Strate, L. (2006). *Echoes and reflections: On media ecology as a field of study.* Hampton Press.

Suzina, A. C., & Tufte, T. (2020). Freire's vision of development and social change: Past experiences, present challenges and perspectives for the future. *International Communication Gazette*, *82*(5), 411–424. https://doi.org/10.1177/1748048520943692

Tichenor, P. J., Donohue, G. A., & Olien, C. N. (1970). Mass media flow and differential growth in knowledge. *Public Opinion Quarterly*, *34*(2), 159–170.

Trench, B. (2006). Science communication and citizen science: Is the deficit model dead? Paper presented at the 9th International Conference on the Public Communication of Science and Technology, Seoul, South Korea, 17–19 May 2006. Retrieved from http://www.pcst2006.org/ Upload/WB1.PDF.

Tufte, T., Jiménez-Martínez, C., & Suzina, A. C. (2020). De-constructing participatory communication and civil society development in 2020: A perspective inspired by Paulo Freire. *COMMONS*, *9*(2), 16–78. https://doi.org/10.25267/commons.2020.v9.i2.01

Turney, J. (1996). Public understanding of science. *The Lancet*, *347*(9008), 1087–1090.

Varona Domínguez, F. (2020). Ideas educacionales de Paulo Freire. Reflexiones desde la educación superior. *Medisur*, *18*(2), 233–243. http://scielo.sld.cu/scielo.php?pid=S1727-897X2020000200233&script=sci_arttext&tlng=en

Vásquez-Guevara, D. H. (2019). *Science for all: Exploring science communication for public engagement in culturally diverse scenarios in the Americas* (Publication No. 13812526) [Doctoral dissertation, The University of New Mexico].

Wagenknecht, K., Woods, T., Nold, C., Rüfenacht, S., Voight-Heucke, S., Caplan, A., . . . & Vohland, K. (2021). A question of dialogue? Reflections on how citizen science can enhance communication between science and society. *Journal of Science Communication*, *20*(3).

Weiss, D., White, J. M., Stohr, R. A., & Willis, M. (2015). Influencing healthcare policy: Implications of legislators' information source preferences for public relations practitioners and public information officers. *Online Journal of Communication and Media Technologies*, *5*(1), 114–135.

Whalen, M. D., & Tobin, M. F. (1980). Periodicals and the popularization of science in America, 1860–1910. *Journal of American Culture*, *3*(1), 195–203.

White, J. M. (2007). *Source choice in agricultural news coverage: Impacts of reporter specialization and newspaper location, ownership, and circulation.* [Doctoral dissertation, Texas A&M University]. OAKTrust Digital Repository. http://oaktrust.library.tamu.edu/bitstream/handle/1969.1/ETD-TAMU-1033/WHITE-DISSERTATION.pdf?sequence=1

White, J. M. (2012a). The communicative action of journalists and public information officers: Habermas revisited. *Journalism Practice*, *6*(4), 563–580.

White, J. M. (2012b). Sabotaging public engagement with science: Missing scientific principles in newspaper stories about the Wakefield MMR-autism

controversy. *Romanian Journal of Journalism and Communication/Revista Romana de Jurnalism si Comunicare, 3–4,* 79–93.
White, J. M. (2014). Influencing policy decisions: Impact of legislator attitudes and demographics on public relations practice. *Online Journal of Communication and Media Technologies, 4*(1), 83–104.
White, J. M. (Executive Producer), Garcia, A. (Director). (2014). *No place like Martineztown* [Film]. Independent.
White, J. M. (Executive Producer), Garcia, A. (Director), & Flynn, C. (Script Writer). (2012). *Just call me Joe.* University of New Mexico.
White, J. M., Morales, G. I., Bentley, J., & Weiss, D. Talking to undergraduates about binge drinking: What channels should we use? Under review.
White, J. M., & Rutherford, T. (2012). Impact of newspaper characteristics on reporters' agricultural crisis stories: Productivity, story length, and source selection. *Journal of Applied Communications, 96*(3), 89–804.
White, J., Willis, M., & Stohr, R. (2013). Legislators' reliance on mass media as information sources: Implications for symmetrical communication between public information officers, public relations practitioners and policymakers. *PRism, 9*(1), 3–20.
White, J., & Wingenbach, G. (2013). Potential barriers to mass media coverage of health issues: Differences between public information officers and journalists regarding beliefs central to professional behaviors. *Journal of Public Relations Research, 25*(2), 123–140.
Wilsdon, J., Wynne, B., & Stilgoe, J. (2005). *The public value of science*. Demos.
Wynne, B. (2014). Further disorientation in the hall of mirrors. *Public Understanding of Science, 23*(1), 60–70.

Chapter 2

Decolonizing Science Communication

A Road for Envisioning Public Engagement

Denisse Vásquez-Guevara, David Weiss, Judith McIntosh-White, and Angélica Ochoa-Avilés

As evidenced in chapter 1, science communication requires evolving from a deficit paradigm to dialogic and participatory frameworks in order to promote newer practices to overcome several limitations that maintain a distant and tense atmosphere between science and society. This chapter presents several reflections that combine decolonial thought and science communication processes to explain how and why research processes have contributed to enacting practices that leave people behind. Moreover, we will offer a reflection on how to practice self-reflexivity applied to science communication studies and tangible practices to conduct research that provides participants a voice. Finally, we will combine decolonial epistemologies with other streams of Western scholarship that provide clear guidelines for researchers who want to develop ethical science communication.

WHY DOES THE SCIENCE-SOCIETY SEPARATION PERSIST?

Research structures and higher education frameworks where knowledge production takes place have followed Western procedures, theories, and methodologies that have dramatically impacted how science is developed as well as the roles researchers and non-expert people have in inquiry. Under Western traditions for science, people have frequently been called *human subjects of study.* Said term has served to analyze, dissect, and extract knowledge in research procedures conducted by researchers and experts in different disciplines, without giving people much agency in the research process. Moreover, Western practices have contributed to creating different portrayals and assumptions when addressing minorities, populations of vulnerable groups, and indigenous groups.

Historically, Western research has been linked to European imperialism and colonization processes of countries in the Global South. Through colonization, academia, and knowledge creation carried by Western nations in the Americas, Australia, New Zealand, and other oriental regions of the world were developed through Western scholarship and frameworks. These studies extracted, collected, and dissected data and take it back to the West, without giving any retribution to the place of its origins. Moreover, sociological studies of knowledge production argue that the West has imposed its own research frameworks for knowledge production under the assumption that knowledge is global and transnational (Boidin et al., 2012). Instead, it perpetuates a system that privileges eurocentric views through conducting research only through frameworks of the West (Krause, 2016). Consequently, Western theories, epistemologies, and methodologies have been used for centuries to produce knowledge not only from own nations but also knowledge from other regions and cultures, excluding other people's voices and cultures.

In the Americas, the disciplines of history, education, knowledge production, geography, economy, and health studies have followed the Western frameworks for centuries to classify and validate knowledge, as part of the many colonization missions of hegemonic power to dominate and produce knowledge of the new world (Martin-Barbero, 1993). Therefore, the way scientists from the Americas have learned to produce and communicate scientific knowledge mostly is conducted from Western research frameworks. Thus, frequent issues such as involving people as mere informants or using Western theories and methods in inquiry that are foreign to the contextual reality in different sciences are reproduced over and over.

Consequently, research has followed a damage-centered approach for nations in the Americas (Tuck, 2009), which works by portraying people as victims of the aftermath of social issues, climate change, and health studies that present detailed descriptions of the brokenness, and, inequality people have suffered, but not their voices and even less their needs and proposals to overcome such issues. However, it is crucial to acknowledge that studies developed under scientific rigor must present evidence of the reality that certain issues create in order to propose possible solutions for them. Still, Western theory-based studies tend to over-focus on the people's vulnerabilities and to narrate the history from a Western-colonial power approach enacted through naming, observing, and extracting knowledge to later be appropriated and commodified (Smith, 1999).

These one-sided portrayals have become a frequent practice giving place to a damage-centered approach which as a research practice might tend to victimize people repeatedly if it's not well balanced with their voices (Denzin & Lincoln, 2014). Decolonial theories and epistemologies depart from deconstructing and criticizing Western research traditions that contribute to perpetuating a framework of revictimization of people. Instead, they provide a framework that focuses on overcoming the social issues affecting people (Mutua & Swadener, 2004).

In the last decade, decolonial theories and methodologies have become an alternative to address issues in the social sciences that derive from the adaptations of Western knowledge. Research developed in the Americas—North, Central, and South America—inherited the epistemologies, theories, and methodologies for knowledge creation of the European countries that colonized their nations (Barbero et al., 2006). Consequently, the local knowledge and culture of Indigenous nations and their people were displaced and replaced along with other acculturation processes, and the emergence of new races and new cultural identities emerged (Denzin and Lincoln, 2014; Jojola and Shirley, 2017), and displaced local knowledge under the assumption of its inferiority to be replaced with Western knowledge, under the argument of scientific rigor and validity.

By the twentieth century, Western research models portrayed people as research subjects. In particular, Indigenous peoples and minorities were portrayed as "dirty," "poor," and "broken," leading to their oppression and silencing (Tuck, 2009). Consequently, these regressive research portrayals became common across the globe. The outcomes of these studies served to reinforce the common history of colonization, oppression, and discrimination through "otherizing" people (Freire, 1970; Martin-Barbero, 1993; Smith, 1999).

For science communication, a discipline currently aims to explore how to overcome issues that derive from the extended use of Western theories, decolonial epistemologies can provide useful knowledge to address its persistent issues from different paradigms that are not only inclusive for target audiences but also enhance and promote their participation in knowledge creation and collaboration with scientists.

What are Decolonial Theories and Epistemologies?

After understanding how decolonial epistemologies emerged, it is important to understand that decolonial thought is a discipline that has many bodies of theory and methodologies. Decolonial epistemologies emerge as a research movement from Latin America, and other Global South native communities as a countermovement to coloniality and hegemonic power through the strong influence of critical theory (Grosfoguel, 2007; Mignolo, 2021) to counterargue and critique the effects that Western knowledge creates.

Decolonial thought can be defined as an epistemic stream that proposes its own theories and methods through different frameworks to overcome power structures inherited from coloniality and to reverse the equation by making Western systems of knowledge the object of critical analysis. Moreover, decolonial studies claim to include the voices and promote the participation of society and researchers (Denzin et al., 2008).

As authors of this book, we do not advocate invalidating all Western theories or demonizing its models and outcomes. Alternatively, we focus on offering researchers a critical analysis of how to open to alternative theories, epistemologies, and methodological approaches to conduct research and implement practical initiatives of science communication. In the following section, we will explore how decolonial frameworks can offer new opportunities to overcome persistent science communication issues.

THE PENDING GOAL OF SCIENCE COMMUNICATION: TO LISTEN AND ENGAGE SOCIETY

Citizen groups and the broader public claim that scientists, universities, and research institutions must engage more with the needs that affect society than only focus on scientific disciplines' research agendas that only benefit academic careers and institutional reputation (Cooper, 2016). Moreover, people need the scientific community to work together to find solutions to the issues that affect them through scientific knowledge. In order to address society's needs, academia, scientists, and research institutions around the globe have developed mostly two initiatives. First, public outreach divisions create and execute projects to serve society through their knowledge. Outreach initiatives reunite the joint efforts of faculty/scientists and who students put into practice their learned skills. As a result, outreach initiatives have been beneficial for communities where scientific or educational organizations are located. Still, these programs or projects are temporary and have simple communication strategies, frequently because of their limited budgets. Second, science communication initiatives, as explained in prior chapters, most frequently follow a deficit-based framework that disseminates information focused on spreading science content through mass media and other conduits, without room for interaction between scientists and people. Consequently, people lack interest in science communication because there is little room for interchange.

Specifically, for science communication, the deficit limitations remain to date through the portrayals of people as incapable of understanding science and their lack of participation and interest in scientific studies. Science communication as a multidiscipline has also privileged Western knowledge and its frameworks. Consequently, as seen in chapter 1, especially through its deficit-based theories, Western theories have influenced science communication research and have led to the scholarly perception that laypeople are only passive recipients of information, lacking the opportunity to give their opinion, critique, or answer back to scientists (Holliman & Whitelegg, 2008).

Therefore, science communication has two pending questions to be answered: How do we overcome the broad public's lack of participation in science? How do we bring the scientific community closer to society and engage it around topics of common interest?

Rethinking Science-Society Relations: Persistent Challenges on Science Communication for Public Engagement

As explained in the first section of this chapter, science and society have a long-time history of separation and lack of interaction. As we had reviewed, some of the reasons are the theoretical frameworks and Western paradigms that guide academia and research. In this section, we will explain the unsolved issues of science communication for public engagement, and the studies that document the consequences and tensions in science-society relations. Moreover, we also aim to create more understanding about the origins and roots of science communication issues that are derived from deficit practices, and which are the initiatives that scientists and faculty started as efforts to improve science and society relations.

Assuming That People Cannot Understand Science

As Stilgoe and Wilsdon (2014) explained, the deficit model approach has guided prior studies in science communication through the notions of public understanding of science, in which researchers' one-sided assumptions argued that people did not know enough and had few skills to understand scientific topics. This approach does not encourage an egalitarian relationship between scientists and people, because it portrays people as ignorant and science as an unchanging matter (Wynne et al., 1995). This assumption evidence how scientists usually perceive the public, which could also be attributed to the lack of training graduate students with master's and doctoral degrees receive to acquire science communication knowledge and skills to approach non-experts (Leach et al., 2008).

Other origins of the assumptions that affect relationships between scientists and people are closely related to the hierarchies inherited from colonial structures and Western knowledge, which influenced educational systems—including higher education—and scientific knowledge production (Hurtado, 2007). Academic hierarchies imposed Western knowledge on their own countries and their colonized nations. Therefore, Western knowledge hierarchies were imposed in twofold academic hierarchies. One hierarchical level imposed the exclusive use of Western theories and methods for inquiry. Another hierarchical level is the structure of privilege that positioned White people, and White scientists as experts whose knowledge is more valuable than the knowledge of other racial groups, in the case of North American nations (Grosfoguel, 2008).

In the case of Central and South America, Latinx[1] also had new privilege groups, the White Creoles and White Mestizos.[2] Therefore, Latinos from Central and South American nations found themselves in a segregated society divided by social stratification and racial discrimination. Moreover, the White mestizo privilege was reproduced in different sectors of Latin American society, such as government and academia, where the indigenous minorities and low-income groups continue to be targets of discrimination and have little or no access to education, healthcare, or representation in politics (Walsh, 2009).

Gender and Privilege Spheres and "The Missing Bodies" in Academia

For centuries the supremacy of privileged groups in the Americas were also reproduced in academia. To date, tenure-track and tenured faculty in universities are mostly White males in the United States, and White male mestizos in Latin America (Bard Wigdor & Artazo, 2017; Ledin et al., 2007). Nowadays, scientific research is still dominated by these groups. Consequently, women and minorities, also known as "the missing bodies," have a small representation in academia. Knowledge production, scientific research, and colonial structures become evident in practices such as the self-granted power to speak on behalf of others under the disguise of benign paternalism (Moreira & Diversi, 2011). As minorities and women are social groups that are frequently affected by different social issues, fewer scholars represent them in academia.

More frequently, social issues are studied by privileged scientists who, due to their social condition, have not experienced such issues as health disparities, gender violence, and inequities, and thus might overlook people's vulnerabilities. Simultaneously, since the research frameworks used have been developed through Western knowledge, they reinforce and reproduce practices that can revictimize indigenous groups, women and minorities, and societal groups as research human subjects who are portrayed as victims of brokenness, powerlessness, and ignorance (Swadener, 2004). The outcomes of these studies, which have exclusively used Western theories and analytical designs, reduce research to simplistic findings that describe results that involve human subjects, their issues, their causes, and their consequences leading to othering people (Smith, 1999). Even more problematic is that research findings do not provide any benefit to society and the research participants, while scientists benefit themselves with publications under their authorship, but instead portray study participants as poor, ignorant, and broken, instead of offering guidelines and pathways to overcome social issues that affect them (Tuck, 2009).

As can be seen, the relationship between science and society requires overcoming colonial frameworks that do not serve human development or help solve the issues that affect society. However, we do not want to assume that all scientists and institutions mostly focus on extracting benefits and knowledge. Nonetheless, it is a reality that university faculty, and scientists in research institutions who are aware of this issue and desire to create research and outreach initiatives to contribute to public service through science, still face the challenges of academic structures developed under Western frameworks. Around the globe, the majority of universities and research institutions still focus their evaluation criteria on academic/scientific production; hiring and promotion still focus on academic work connected to reporting, teaching, and obtaining grants in a bureaucratic system that is highly time-consuming, instead of prioritizing collaborative work that reunites science and society. This phenomenon is present in universities and research institutions and is better known as "publish or perish" in the academic and scientific community."

Barriers to Science-Society Relations and Standards of Universities and Research Institutes

Research is one of the main and most valuable practices of universities and scientific organizations. Higher education institutions face a highly competitive scenario where universities are ranked based on their ability to get funding through research grants, and the number of highly ranked publications their faculty produce. Moreover, these indicators and data are compared to other universities in national and international ranks that reinforce their institutional prestige.

For graduate students and faculty, higher education demands research functions as well as teaching activities as part of their academic duties/workloads, along with teaching. Primarily, the number of publications and grants awarded for research projects are some of the most valuable attributes researchers can have in their curriculum vitae. Moreover, those elements are crucial efficiency indicators that are reviewed during faculty searches and student applications for doctoral programs, but also for faculty promotions by university committees.

Therefore, the relationship between science and society frequently becomes secondary as the competitive nature of academia leaves few spaces and little time for researchers to engage with the public. Consequently, science communication also falls into secondary service requirements in research institutions. Moreover, the requirements of universities, funding agencies, academic colleagues, and scientific associations that direct scientific dissemination does not provide space and funding for science communication and public engagement (Stilgoe et al., 2014). Still, the expected outcomes of research are attending scientific conferences and publishing in academic journals. However, studies developed in the United Kingdom that explored scientists' willingness and interest in public engagement and determined that they would like to spend more time sharing the results of their studies through dialogue; however, the academic system and the duties imposed on them to maintain their jobs or get promotions are the barrier for this to happen (*NSF*, 2004; The Royal Society, 2017).

However, over the past decade, several studies have found that scientists and faculty in the United States were becoming more interested in exploring ways to become more engaged with society, even when they faced multiple barriers and even threats to their careers. In summary, studies evidence that the academic environment in the United States and other parts of the world have changed over the past decades. Scientists do feel identified and acknowledge that engaging with the public more fully in their research is necessary (Campos, 2022; Canfield et al., 2020; Lane, 1997).

Science and the Barriers for Public Engagement: Issues and Countermovements

Moreover, scientists, faculty, and students can become change agents by becoming more engaged and conducting research about the real needs of their communities and inviting people to collaborate with them. Consequently, public engagement and science communication are a way to advance the common good by exercising the civic duty of contributing through expert knowledge to the issues that affect society (Greenwood & Riordan, 2001; Nelson & Vucetich, 2009).

Nonetheless, in this current time, amid the politically polarized scenario, this shift in the roles of scientists in academia and society has become a major risk for scientists, because these efforts could jeopardize their job stability, scientific reputation, and even personal safety. During the Trump administration (2017–2020), the scientific community faced multiple struggles, which also represented multiple barriers for scientists and science communication. However, in other parts of the world, attacks and harassment to scientists from politicians and ideological movements are not new (Kling, 1963). In Ecuador, academics and scholars who advocated for minorities, and others who participated in corruption investigations were fired from their jobs, incarcerated, and sought political asylum (Becker, 2013).

As a response to political barriers, the global movement *March for Science* is an example that there is a significant number of scientists interested in advocating for science and how it should be more involved with society by engaging research as a source of information and analysis in policymaking processes (Ross et al., 2018). As a movement, March for Science has emerged primarily as an advocacy group with members of the scientific community that rejected the budget cuts that the Trump administration executed to federal science agencies that provide research funding (Mooney, 2017).

Another policy restraint, more specifically related to science communication and the barriers that the scientific community faces, were the governmental policies developed by the Trump administration that limit how scientists can disseminate their research to the public. These policies were imposed in 2017 to prohibit scientists and professors from presenting their research results to society at large in events or mass media (Lin, 2019). This policy represented a restriction of scientists' constitutional rights as citizens to speak and present their knowledge, and to express themselves freely.

During the Trump administration, scientists in the United States were not protected by law if they developed any kind of scientific communication; this issue leads to self-censorship and suppression if the research discipline addresses political or economic power groups (Rainey & Rainey, 2017). Therefore, from 2017 to 2020, the scenario in the United States for science communication became a risky practice for researchers. During these years, many scientists avoided public science communication for fear of affecting their work stability as well as their role in society. Nonetheless, still many scientists organized to protest against this restrictive law and advocated to defend their right to be engaged with society through their scientific studies.

By 2021, many scientists advocated for the White House Office of Science and Technology Policy (OSTP), the Centers for Disease Control and Prevention (CDC), and the US Environmental Protection Agency (EPA) initiated dialogue with many scientific grassroots organizations and advocacy groups from universities to abolish any restrictive laws against to share scientific information to society and motivate public engagement. These conversations have met with success (Subbaraman, 2021).

At the time of writing this book, an important number of the science community across many countries are advocating to bring science and society closer in collaboration (Campos, 2022).

ACADEMIC LIFE: HOW TO DISMANTLE THE DEHUMANIZATION IN SCIENCE COMMUNICATION THROUGH DECOLONIAL THOUGHT

Decolonial epistemologies from education and cultural studies suggest several principles that can inform how to attend to the persistent issues in science communication for public engagement. This section will offer a reflection on how to address—in research and practice—the issues explained in the prior sections in science communication through several studies of decolonial scholars with a multidisciplinary approach.

Self-Reflexivity Practices for Science Communication Projects and Professionals

After describing the limitations academics and researchers face when they aim to engage with people by conducting research or implementing initiatives for science communication, and how these emerged, we propose several principles of decolonial principles. These principles are meant to guide and inform self-reflexivity practices that attend to the reproduced practices that reproduce the disengagement of audiences.

Overcoming the Privilege Frameworks

As academic and scientific structures have mostly worked through hierarchies to understand researchers' and people's positionality, science communication requires us to reflect critically on how to overcome barriers to public engagement. bell hooks, an American feminist educator, in her book *Teaching to Transgress,* proposes a pathway to rethink democratic participation. In her work, hooks advocates the analysis of the field of critical pedagogy to address the issue of educators who do not want to teach, and students who do not want to learn, and how the issues of racism and sexism in the classroom affect students' and instructors' interests. Through these notions, science communication can explore the issues of the lack of mutual willingness for collaboration among scientists and people. These principles originate from the notion of transgression, as a practice to overcome existent boundaries that limit interaction and interchange. Such boundaries frequently are, but not exclusively: social sectarianism and discrimination, inequalities in the access to resources and opportunities, and the wrongful use of the privilege by people in power. Specifically, for science communication, transgression from a decolonial standpoint means reflecting on how to develop inclusive conditions for collaboration between scientists and people. Collaborations in science-society must seek to address social issues that intertwine racial, gender violence, inequalities, and health, among others.

In order to transgress the limitations of the relationship between scientists and the public from a social change standpoint, the concepts of *service of love* as part of *love as a radical practice provide* us with relational politics guidelines to approach the ethics to practice and manage relations between scientists and the publics (hooks, 2014; Nussbaum, 1990).

Service to Others

This principle explains a starting point to begin understanding people collectively. In order to start referring to "we" instead of the "otherizing" tradition of "them," the best approach is to engage in thinking that we are in service to each other (Porter, 2010). In the context of science communication, an alternative starting point is to understand that researchers and people are equals who construct and help each other to build knowledge and actions to improve and learn together. Moreover, is important to acknowledge that scientists contribute from their knowledge and training. Alternatively, people can contribute from their lived experiences and practice. Therefore, no knowledge is superior but rather complementary to each other.

Compassion

As a decolonial principle, compassion is a practice of self-awareness regarding the impact that our actions might have on the world (Porter, 2010). Consequently, for science communication, compassion implies reflecting on how people will be impacted by scientific content before we initiate science communication and public engagement activities.

Compassion can be put into practice through the processes of practicing another decolonial principle, which is cultural humility.

Cultural Humility

Several researchers conceptualize cultural humility as a proactive practice to recreate awareness about people´s reality, needs, and vulnerabilities before research teams contact them (Hacker, 2013). Incorporating cultural humility principles as ethical considerations in research and practice enriches researchers' self-reflexivity, addresses their cultural beliefs, and counters power imbalances that can emerge with research participants (Foronda et al., 2016; Tervalon & Murray-García, 1998). Some useful considerations to incorporate cultural humility are the following:

- *Engaging Formative Research.* Formative research is a useful tool to create awareness about the cultural characteristics, identity, and any sociodemographic characteristics (i.e., poverty, vulnerabilities, and any other limitations in access to information and participation regarding a specific project topic). Formative research also helps researchers to be well informed about the audiences, before contacting them for developing an inquiry or a project. Some guidelines for developing engaging science communication emphasize the need to develop formative studies to better understand the audience's use of media platforms that can facilitate conversations (Nisbet & Scheufele, 2009).
- *Demonstrate True Willingness for Meaningful Collaborations.* Sadly, researchers frequently missed science communication for personal promotion strategies. These strategies center their attention on the scientists' skills and credentials. Consequently, science content takes a secondary role and loses, its true value to people. To avoid this issue, we suggest that researchers engage in cultural humility to reflect on how our research or practical initiative will share power with people. For example, it would be important to locate grassroots organizations, associations, or volunteers actively interested and involved in the main topic of the project. For instance, if our science communication initiative focuses on food safety practices of street food vendors, researchers and communication specialists should locate the key groups that congregate food vendors to invite them to dialogue about the issues they face in their businesses, and which are their unsolved issues, interests, and needs. A second step would be to invite vendors to collaboratively plan a research project or a science communication initiative, include them in the decision-making process, and develop/identify a collaborative setting with shared responsibilities.

  Another strategy to demonstrate willingness is to create opportunities to train non-expert people on skills or procedures that can help them to address their issues.
- *Negotiate the Agenda and Focus of Science Communication.* Whether the project implies research or practical initiatives, meaningful cultural humility and compassion are meant to prioritize and maximize the audiences' benefits. Therefore, scientists who are driven by ethical goals focus on solving and addressing issues that affect people, and less on gaining public attention, with a goal of gaining fame or public recognition (Vásquez-Guevara, 2019).

Self-Reflective Practices: How to Create Reflective Processes for Science Communication

After summarizing some principles and suggested practices to implement ethical considerations that ensure inclusive and respectful procedures, it is important to explore how to implement self-reflexivity in science communication work teams or research groups. Reflective practices is a research procedure more frequently found in qualitative research "where it is used to legitimate and validate research procedures. Today, the authoritativeness of reflexivity is practiced in a wide range of research schools: critical, feminist, race-based, and post-structural approaches" (Mortari, 2015).

Additionally, self-reflexivity is crucial to develop protocols that guide researchers' behavior specifically focused on how to address research participants and manage their positionality, and privilege, balancing their power and agency in the research process. Learning and practicing self-reflexivity is a foundational practice for researchers who work with people in their inquiry endeavors. It allows people to engage in thoughtful relationships with the setting they are going to intervene in and to gain awareness about the impact of their work in people's lives and knowledge creation (Boud et al., 2013; Mortari, 2015).

In science communication, self-reflectivity is based on the need for people to consider the aims and outcomes of their research. Self-reflectivity exercises imply that scientists describe their research motivations

> in terms of particular theories or particular localized research problems, they are often also motivated by broader social aims—curing cancer, addressing energy sustainability, finding alternatives to expensive or impractical approaches that may be theoretically "clean" but practically complex. The mental exercise of reflection is very important for placing one's story in the greater picture and making it relevant to societal concerns. Reflecting on the outcomes of one's research is particularly important since this is what many audiences would consider most important. Scientists tend to complain about the public's preference for "jumping ahead" to the implications of basic research, but the literature on science communication makes it clear that scientists would be wise to consider the implications as well. (Baram-Tsabari & Lewenstein, 2017)

Some tools to develop self-reflexivity consider the use of self-reflection memos, discussion sessions, and the development of procedures to manage power and decision-making with others. These practices can be developed individually or through collective sessions, combining individual and private research exercises and collectively constructed reflections.

In case study 1 we present how we incorporated self-reflexivity practices adapted to a science communication study. *Science for All* required that researchers and former participants of health promotion studies be united to evaluate and co-design solutions to better engage their target audiences in two different scenarios. We describe how self-reflexivity was applied as a collective practice to reflect on how to foster equal participation and mutual learning.

Case Study 1: Exploring How to Improve the Audience's Engagement in Health Promotion Programs for Children

The *Science for All* study focused on exploring science communication for public engagement in two health promotion programs in Ecuador and the United States. The goal of this study was to explore how to improve the opportunities to engage young audiences and their parents around two health promotion programs: Eat Smart to Play Hard (UNM Prevention Research Center in Albuquerque, New Mexico, US) and ACTIVITAL (Biosciences Department at the University of Cuenca, in Cuenca, Ecuador). As the study was developed through the methodology of Participatory Action Research (PAR), it was crucial to design procedures that would not attend exclusively to the researcher´s interests but provide voice and agency to the audience in order to understand their opinions, needs, and interests around appropriate nutrition and physical activity. Four co-design workshops were conducted with each health promotion program group.

Self-reflexivity sessions were incorporated by using the principles of cultural humility. The study was designed as a collaboration between researchers and young audiences. As the PAR framework promotes egalitarian agency and audience participation, researchers were required to acknowledge their privilege and power and how they would position themselves in relation to the co-researchers, the audience. One of the researchers of this study facilitated a one-hour reflexivity session with five of the researchers of ESPH in Albuquerque, and later separately with two of the researchers of ACTIVITAL in Cuenca. Before the session, the researchers were asked to bring a clear written statement acknowledging their privilege in terms of socioeconomic characteristics and describing their identifications in terms of gender, race, culture, and ideology. At the session, the researchers were asked to disclose how they would use the emerging data to ensure mutually beneficial outcomes for the co-researchers and their own programs, focusing not only on scientific outcomes but also on activities that would provide voice and agency to the co-researchers in the co-design process.

As a result, researchers agreed that the first of the four co-design workshops would have dedicated time to set the rules with the co-researchers to enhance egalitarian agency and decision-making.

Therefore, decolonial theories were applied as guidelines to design and propose research procedures that create an egalitarian setting among researchers and participants. After researchers developed self-reflexivity exercises guided by the principles of cultural humility (Foronda et al., 2016; Hacker, 2013), the following guidelines were included in the research protocol:

- Participants would choose the locations, dates, and times to conduct the participatory workshop sessions, in places that are accessible safe, and comfortable for them.
- Participants and researchers of each case study would construct the final ethical procedures and policies to proceed in the research sessions.
- At all times, the participants would be invited to co-facilitate the research sessions and formulate additional questions.
- Participants would be invited to lead the co-design workshops.

Building Collaboration and Participation

Decolonial thought claims to address power imbalances that frequently emerge in research processes. The third step, after self-reflexivity and approaching people respectfully to build partnerships and collaboration in research is crucial to design processes that are participatory and provide an egalitarian structure for collecting data and developing initiatives that address the interests and needs of people and researchers.

The principles of decolonial epistemologies can help determine the procedures for developing dialogical encounters and explain how shared concerns about social issues can foster a sustainable relationship in immediate, medium, and long-term projects among different social groups (Smith, 1999). These principles can bring ethical considerations to bear on integrating cultural and racially diverse groups such as mestizos and indigenous groups in Latin America into the process of research (Walsh, 2016). The United States, as a culturally diverse country, could similarly transform the practices in academia and consider these decolonial principles (Porter, 2010). Moreover, these principles can be used as guidelines to create tangible practices based on spaces to open dialogue among people of different social backgrounds as researchers and the public.

This section will explore how the principle of insight and participatory action research (PAR) can be useful to translate decolonial principles into practice.

Insight

As a principle, insight provides guidelines to solve emergent issues of the group's claims and dilemmas that can appear in planning processes based on the context. Here, useful reflections can occur to foresee the possible difficulties in the research process and to develop actions to consider strategies oriented to facilitate the emerging issues (Porter, 2010).

Science communication scholars, in the last two decades, advocate for the development of research and practices through participation paradigms in order to overcome deficit frameworks and develop a further exploration of audiences' needs and interests for science communication (Davies, 2008; Hackett et al., 2008; Van Even & Zaman, 2020). Newer studies that applied participatory action research methods along with participatory data collection tools provided valuable evidence for studies of different disciplines to update or design from the start initiatives that are more engaging to their audiences (Rubin, 2021; Trench, 2021).

Participatory Action Research

Drawing from participatory communication and emancipatory pedagogies from Paulo Freire, a new approach to organizing models is based on integrating and organizing people around a topic or issue of interest. The participatory models are based on radical participatory practices rather than on representational democracy, better known as a revolutionary vanguard model (Melling & Pilkington, 2018). This model reconceptualizes "safe space" as a political movement that flows from the logic of privilege regarding gender, race, and class to a different egalitarian framework for participation (Fernández-Aballí Altamirano, 2020). PAR methodologies will be further explained in chapter 3.

RETHINKING THE ROLE OF SCIENCE COMMUNICATION FOR PRACTITIONERS AND RESEARCHERS

Currently, science communication faces multiple challenges that need attention and effort from researchers and practitioners. This section will present several considerations for both groups to address two current issues that are priorities on science communication research agendas. These issues must be considered for research and practice, for audiences' inclusion and participation, and for how scientists can and should participate in the public sphere by providing scientific data that gives useful information to the broad public and policymakers.

Inclusive Science Communication

There is a strong call to improve and explore how to practice inclusiveness in science communication, derived from the issues created by the deficit model. Studies frequently refer to the issues created by the lack of inclusiveness in science communication as barriers (Dawson, 2018). These barriers refer to two factors, (1) the target audience's demographic conditions, its geographical location, and its access to science centers, institutions, or universities where events take place, and (2) the lack of audience's interest in some science topics, which does not motivate people to participate in science communication activities or be informed about specific topics. This audience group is also named by scholars as *disengaged audiences* (Burns & Medvecky, 2018). Science communication has broadly explored how to address disengaged audiences and attract their interest in science content, and how to engage them around science topics that affect people's daily lives (Cormick & Romanach, 2014; Gregory & Miller, 1998).

However, there are only a few studies that address inclusiveness in science communication (Canfield et al., 2020; Humm & Schrögel, 2020), although some meaningful contributions have been made from the cultural studies approach. The findings of these studies evidence that there are underlying tensions involved in developing science communication initiatives based on a barriers perspective that underestimates the underlying issues of social exclusion in science communication (Dawson, 2018). Consequently, social exclusion is a persistent issue that intersects through multiple inequalities that audiences face in their lives, such as power resting on researchers in the processes of science communication practices, the limited access to educational centers or resources, or tools such as the Internet, that enable audience participation. Therefore, the unattended issues of inclusion in science communication are complex, because these respond also to the structural inequalities of vulnerable audiences and populations.

Consequently, when science communication aims to propose inclusive initiatives or research, the barriers approach is not recommendable, because it might perpetuate the exclusion of vulnerable audiences. After all, researchers and practitioners might have limited power to solve these issues (Canfield et al., 2020; Levitas, 2005).

A future challenge for science communication researchers, practitioners, grant agencies, and policymakers is to develop solutions for inclusiveness in science communication. One way scholars and practitioners suggest to overcome issues of social exclusion is to explore one-off inclusive projects. These one-off initiatives offer valuable opportunities that are open to people through free events or activities, which can take place at public spaces, such as parks, museums, and schools. Additionally, another set of strategies to overcome issues of social exclusion is to diversify scientists' representation by developing events and activities that show minority-group scientists (i.e., women of color, Latinx). These types of events have evidenced that they are efficient in providing cues for social belonging and identification with audiences of historically excluded groups (Canfield et al., 2020).

Additionally, as social exclusion itself is an issue that intersects the physical, cognitive, and social dimensions, science communication can use these criteria to reflect how vulnerable audiences can be included in order to design inclusive science communication initiatives (Reich et al., 2010). Several practitioners have also explored arts-based initiatives and other strategies to motivate inclusion and audience engagement that have not been peer-reviewed, such as the use of music to communicate scientific findings of animal behavior and other natural sciences (Lee et al., 2019); however, more studies are still required.

Scientists as Public Figures, or Science as the Main Focus

Science plays a key role in society´s development and provides data to guide decision-making in health, education, economy, law, and many other fields that affect large populations. Scientists throughout history have opted to speak publicly about the findings of the studies' findings, while at the same time proposing alternatives to solve issues that affecting society. Some scientists have even become worldwide public figures that motivated young people to follow their legacy and work (Krauss, 2015).

Albert Einstein, Stephen Hawking, Marie Curie, Carl Sagan, Bill Nye, and most recently Anthony Fauci (with his active role in the COVID-19 pandemic) became public figures of their time, presenting to the world the results of the results and discoveries of their studies, thus raising awareness. All of them became publicly known for their work and contributions to science, their advocacy to include science in the process of political decisions, and most importantly for awakening the interest of society and steering the public opinion around controversial topics such as health and energy resources. Nevertheless, with social media and massive mediatization, current and future scientists need to reflect more on their role as experts, and critically consider crucial questions that need honest answers: *Do I need to be the focus of attention in science communication? Is the message about science the main goal of science communication?*

We advise you to incorporate these questions into your self-reflexivity practices before developing your science communication projects. Reflect about how your personal and professional goals can be balanced in order to put into practice inclusivity and establish a mutually beneficial relationship between you as a scientist/researcher and your audience.

Case Study 2: Dr. Anthony Fauci and Communication amid the COVID-19 Pandemic

A largely public case is the role Dr. Anthony Fauci had in the 2020's COVID-19 pandemic. As head of the National Institute of Allergy and Infectious Diseases (NIAID) (NIAID, n.d.), Fauci faced multiple challenges while he led the COVID-19 Taskforce of the White House under the Trump administration. Some of these challenges included: a tense political polarization, the anti-vaccine group's misinformation, and a collapsed health system facing almost a million deaths since March 2020 to the date this book is written.

As we mentioned before, communicating science is a challenging task, as it implies facing counter arguments from multiple sources that might or might not be merely scientific, but instead could respond to political or ideological discourses. In the midst of social panic and mandatory social distancing, uncertainty grew in the United States, as well as around the world. However, the role scientists and public information officers played, specifically in Fauci's case, was crucial in managing the pandemic through the media by providing prevention guidelines and information, such as social distancing, the use of masks, and later on vaccination campaigns.

Fauci opened multiple spaces for public health communication in mass media such, as television, newspapers, magazines, and social media almost on a daily basis, to inform audiences and respond to common questions from multigenerational audiences. Here, the main goal was to promote risk-reduction messaging strategies to shape and influence the audience's perceptions and attitudes of audiences (Nan & Thompson, 2020) and apply health behavior change models in order to support the efforts that the health system required to lower contagion and spread of COVID-19, and consequently the number of deaths.

In an interview for the Research America 2021 National Health Research Forum titled "Researcher to Researcher: A Fireside Chat with Dr. Anthony Fauci" (2021), the NIAID Director referred to the following guidelines for science communication:

> Well I think there should be some training. I believe since we, as you're alluding to, we really need to get more scientists out there communicating for a number of reasons, one of the most important which is that there is a considerable amount of misinformation and disinformation out there particularly on social media and scientists really should feel comfortable in articulating the facts and the evidence that gets people to make the appropriate decisions when it comes to

> their health. So I'm not so sure it should be a long-drawn-out formal training but there should be some fundamental training. I mean some fundamental principles that I abide by I could teach in one class, you wouldn't have to go through the entire semester of learning, you could probably do it in one class.
>
> One of the most important things is that you don't, if you're a scientist and you're communicating, the objective of what you're trying to do is not to appear smart, it's to be understood. You know, sometimes scientists when they get up in front of people, they feel they want to be as detailed and as arcane as they possibly can, that is a big mistake, people need to understand what you're talking about, they don't need to be impressed by how smart you are, so there are a couple of things like that. Also say in a few words what some people might say in a lot of words, short is better always. (n.d)

As Fauci states, a balanced and self-aware position is not to assume that scientists must show how technical they are, their credentials and experience already speak for themselves. Instead, scientists must be trained to focus on audiences and to create approachable and understandable narratives that are useful and clear enough for audiences to put guidelines into practice.

Fauci was later named "America's Doctor" and became a viral phenomenon on social media. People have frequently admired his approachability and openness without losing academic rigor as well as his role as a scientist above a public figure. Nevertheless, some audiences deeply criticized his appearance in fashion magazines, and outside news or scientific journals. This was the case regarding his appearance on the *InStyle Magazine* cover in 2020 for which he gave a personal interview about his life and family but also touched on crucial implications of COVID-19 prevention based on the available studies by 2020.

If as Fauci states in his interview for Research America 2021 National Health Research Forum, and the guidelines suggested in this chapter, if the main goal is to reach audiences with understandable and useful information, scientists need to leave their comfort zone of traditional strategies to reach audiences with key information that can affect their lives. The massive spread of COVID-19 also needed massive dissemination of prevention guidelines, which in the age of media specialization sometimes just does not reach everyone. In our own criteria, the *InStyle Magazine* piece about Fauci not only helped to reach audiences who sometimes choose not to watch the news with key messages (e.g., the importance of following prevention and infection guidelines), but it also humanized the role of a scientist amid a historical crisis. Additionally, it presented a more balanced and nuanced profile of a scientist and the effort behind developing health research.

Scientists as Public Figures

Science communication started attracting more attention in the twenty-first century, as mass media and social media became the arena for reaching public representation and recognition of science and scientists. Therefore, science communication initiatives also became a tool to position scientists' public images and their projects. However, scientists mostly focused on building a personal reputation that gives them recognition, respect, or even fame as a strategy to position themselves and guarantee, funding to finance their research projects (Daza-Caicedo et al., 2016).

As mass media and new media (social media and internet-based information platforms) evolved dramatically in the last century, media power also transformed society from written-illustrated to visual-entertainment communication. In the case of science communication, some strategies and initiatives also evolved from the logic of motivating audiences to reflect about a topic, to the lab-based researcher to a celebrity of screens or a scientist-politician (Aguirre et al., 2016). Some scholars are critical about scientists focusing efforts on communicating their image or their achievements instead of delivering to society key messages about their knowledge to society to provide openly useful information that can benefit people by helping them make well-informed decisions with science-based information.

The Colombian scholar Omar Rincon (2016) suggested the term *egoteca,* in its English translation egotistical communication, which refers to science communication strategies that have the sole intention of making an audience recognize a scientist´s social value. By using these strategies, science content becomes secondary to the scientist's public figure, which becomes the main goal of communication strategies and tactics.

Hence, science communication research and practical initiatives claim academic rigor and valuable content. The newest science communication research movement, such as science communication for public engagement and technology, PEST, instead demands a focus on how to engage people around science topics, and make them active participants alongside scientists for mutual learning (Bennett & Jennings, 2011).

Nevertheless, it is crucial to recognize that scientists who develop research doubt how to communicate their studies' findings without losing their academic rigor. Several studies reveal that scientists recognize the value of sharing their knowledge and data that can be useful to the broader society (Greenwood & Riordan, 2001; Haywood & Besley, 2014; Krauss, 2015). Other barriers to scientists' willingness to develop science communication initiatives are related to the need to prioritize their efforts to meet the productivity criteria that universities and research institutes impose on them to maintain their jobs. The 2004 NSF's Science and Engineering Indicators

Survey found that 76 percent of researchers said they do not have enough time to communicate with nonexperts, 28 percent do not want to take part in such activities, and 17 percent do not care at all about science communication (NSF, 2004).

Regarding the roles that scientists take in developing science communication practices, newer studies by The Pew Research Center (Rainie et al., 2015) applied to U.S.-based members of the American Association for the Advancement of Science (AAAS), evidenced that "AAAS members were broadly supportive of having scientists contribute to public discussions about scientific issues. Specifically, 87% of AAAS respondents said that scientists should 'take an active role in public policy debates' related to science and technology. In contrast, 13% said that scientists need to 'focus on establishing sound scientific facts and stay out of public policy debates'" (Rainie et al., 2015, p. 820). Consequently, there are significant tensions in researchers' perceptions and attitudes regarding science communication and how its ethical practice should be carried out. Thus, there is a dichotomy between a major group of scientists who argues that participating in policy debates is crucial and a minority that instead claims to limit scientists' participation to deliver scientific data only.

Envisioning public engagement for science communication does require scientists' active involvement, but it is also crucial to propose a framework that embraces academic rigor, the media's ethical use, and the main focus on scientific content above scientists' protagonism. However, it is understandable that scientists want to communicate their studies and contributions under their authorship. Nonetheless, as explained previously, seeking mainly public recognition, and sometimes aiming for celebrity statuses without substantial contributions to science and society is not ethical science communication.

In his work, *The Image: A Guide to Pseudo Events in America* (1992), Daniel Boorstin, an American historian, and author, addresses the phenomenon of heroes versus celebrities. Boorstin argues that the root of the problem of celebrities relies on the modern capacity to make individuals famous through fabricated media efforts. These efforts aim to gain people's admiration and recognition as an effort of self-advertising a person even when they lack qualities that are valuable to audiences (Boorstin, 1992).

Heroes, instead, are admired by people for their human positive values such as courage, kindness, abilities, knowledge, and society's contributions to the common well-being. Heroes are self-made as the result of their efforts during their lives, with successes and failures across time. More importantly, heroes frequently do not seek public recognition as a goal, but instead they are committed to a major purpose, such as advocating for social change, developing alternative health treatments, or proposing technological innovations that benefit society.

Closer to the questions opened in this section, Boorstin's work (1992) explores scientists' role, and how historically they had a crucial role in the areas of human progress and technological development. Scientists have worked as profound thinkers who innovated through working beyond the lay public field's understanding (Boorstin, 1992). In modern times, the presence of scientists in mass media is more and more distant. For example, currently, there is almost no science-focused news programming on the three major broadcast networks in the United States (ABC, CBS, and NBC) (NSF National Science Board, 2016).

Then, what we can borrow this discussion about how to develop ethical science communication in the current time? First, scientists need to reflect that science communication must serve science. Second, their presence as experts and authors is crucial, especially by creating spaces where they can interact and dialogue with society with the purpose to motivate mutual learning and promote public engagement of target audiences and key stakeholders.

CONCLUSION

Decolonial thought creates more understanding to dissect and understand the issues that act as persistent barriers in science communication for public engagement. Moreover, it enables us to reflect on research procedures that can be useful to develop science communication initiatives that overcome disengagement and privilege in order to promote egalitarian research settings, which are mutually beneficial for people and scientists and promote active participation.

Further on in chapters 3 and 4, we will focus on how to incorporate science communication in multidisciplinary research projects around health and the main considerations for funding agencies, but also on how to plan, design, and execute science communication initiatives that are engaging to the target audiences and contribute with valuable knowledge to people.

NOTES

1. Latinx is used as a gender-neutral noun for an individual of Latin American Heritage (Merriam-Webster, n.d.).

2. White Creoles and White Mestizos are the racial groups that emerged in Central and South America after the Spanish colonization. Creoles are Spaniard direct descendants born in Latin America while Mestizos are people born from Whites and Central or South American Indigenous people (Padilla, 2019).

BIBLIOGRAPHY

Aguirre, C., Arboleda Castrillón, T., Cassiani, S., Daza Caicedo, S., Guivant, J., Hermelin Bravo, D., Hilgartner, S., Jassanof, S., Lozano Borda, M., Lozano Hincapié, M., Maldonado Castañeda, O. J., Metcalfe, J., Olivé, L., Raigoso Camelo, C. E., Sutz, J., Tafur Sequera, M., & Ramírez, M. A.. (2016). Ciencia, tecnología y democracia: reflexiones entorno a la apropiación social del conocimiento. http://repositorio.minciencias.gov.co/handle/11146/248

Baram-Tsabari, A., & Lewenstein, B. V. (2017). Science communication training: what are we trying to teach? *International Journal of Science Education, Part B, 7*(3), 285–300. https://doi.org/10.1080/21548455.2017.1303756

Barbero, J. M. (2006). Intervening from and through research practice: Meditations on the Cuzco workshop/31. In D. Sommer (Ed.), *Cultural agency in the Americas* (pp. 31–37). Duke University Press. https://doi.org/10.1215/9780822387480.

Bard Wigdor, G., & Artazo, G. (2017). Pensamiento feminista Latinoamericano: Reflexiones sobre la colonialidad del saber/poder y la sexualidad. *Cultura y Representaciones Sociales, 11*(22), 193–219. http://www.scielo.org.mx/scielo.php?pid=S2007-81102017000100193&script=sci_arttext&tlng=en

Becker, M. (2013). The stormy relations between Rafael Correa and social movements in Ecuador. *Latin American Perspectives, 40*(3), 43–62. https://doi.org/10.1177/0094582X13479305

Bennett, D. J., & Jennings, R. C. (2011). *Successful science communication: Telling it like it is.* Cambridge University Press. https://play.google.com/store/books/details?id=cw98Z78rjZIC

Boidin, C., Cohen, J., & Grosfoguel, R. (2012). Introduction: From university to pluriversity: A decolonial approach to the present crisis of Western universities. *Human Architecture: Journal of the Sociology of Self-Knowledge, 10*(1), 1–6. https://search.proquest.com/openview/04a5cdb2c614f571f0fe7d95a7639859/1?pq-origsite=gscholar&cbl=40294

Boorstin, D. J. (1992). *The image: A guide to pseudo-events in America.* Vintage.

Boud, D., Keogh, R., & Walker, D. (2013). *Reflection: Turning experience into learning.* Routledge. https://play.google.com/store/books/details?id=XuBEAQAAQBAJ

Bucchi, M., & Neresini, F. (2008). Science and public participation. In E. J. Hackett, O. Amsterdamska, J. Wajcman, & M. Lynch (Eds.), *The handbook of science and technology studies* (pp. 449–472). The MIT Press.

Bucchi, M. & Trench, B. (2021). *Routledge handbook of public communication of science and technology: Third edition.* Routledge. https://doi.org/10.4324/9781003039242

Burns, M., & Medvecky, F. (2018). The disengaged in science communication: How not to count audiences and publics. *Public Understanding of Science, 27*(2), 118–130. https://doi.org/10.1177/0963662516678351

Campos, R. (2022). Including younger children in science-related issues using participatory and collaborative strategies: a pilot project on urban biodiversity. *Journal of Science Communication, 21*(2), N07. https://doi.org/10.22323/2.21020807

Canfield, K. N., Matsuda, S. B., Menezes, S., Moore, A., Mosley Austin, A. N., Dewsbury, B. M., Feliú-Mójer, M. I., McDuffie, K. W. B., Moore, K., Reich, C. A., Smith, H. M., & Taylor, C. (2020). Science communication demands a critical approach that centers inclusion, equity, and intersectionality. *Frontiers in Communication*, 5. https://doi.org/10.3389/fcomm.2020.00002

Cooper, C. (2016). *Citizen science*. Overlook Press.

Cormick, C., & Romanach, L. M. (2014). Segmentation studies provide insights to better understanding attitudes towards science and technology. *Trends in Biotechnology, 32*(3), 114–116. https://doi.org/10.1016/j.tibtech.2013.12.005

Davies, S. R. (2008). Constructing communication. *Science Communication, 29*(4), 413–434. https://doi.org/10.1177/1075547008316222

Dawson, E. (2018). Reimagining publics and (non) participation: Exploring exclusion from science communication through the experiences of low-income, minority ethnic groups. *Public Understanding of Science, 27*(7), 772–786. https://doi.org/10.1177/0963662517750072

Daza-Caicedo, S., Lozano-Borda, M., Bueno Castellanos, E. M., Gómez-Morales, Y. J., Salazar Acosta, M., Jaime, A., Aguirre Guzmán, J. P., Rueda Ortiz, R., Franco-Avellaneda, M., Rincón, O., Pérez-Bustos, T., Farías, D., Suárez M., R., Osorio Marulanda, C., & Sistemas Especializados de Información, S. E. I. (2016). Percepciones de las ciencias y las tecnologías en Colombia. Resultados de la III Encuesta Nacional de Percepción Pública de la Ciencia y la Tecnología. http://repositorio.colciencias.gov.co/handle/11146/291

Denzin, N. K., & Lincoln, Y. S. (2014). Introduction: Critical methodologies and indigenous inquiry. In N. K. Denzin, Y. S. Lincoln, & L. T. Smith (Eds.), *Handbook of critical and indigenous methodologies* (pp. 1–20). SAGE Publications, Inc. https://doi.org/10.4135/9781483385686.n1

Denzin, N. K., Lincoln, Y. S., & Smith, L. T. (2008). *Handbook of critical and Indigenous methodologies.* SAGE. https://play.google.com/store/books/details?id=MJ8R41fRT08C

Fernández-Aballí Altamirano, A. (2020). The importance of Paulo Freire to communication for development and social change. In J. Servaes (Ed.), *Handbook of communication for development and social change* (pp. 309–327). Springer Singapore. https://doi.org/10.1007/978-981-15-2014-3_76

Foronda, C., Baptiste, D.-L., Reinholdt, M. M., & Ousman, K. (2016). Cultural humility: A concept analysis. *Journal of Transcultural Nursing: Official Journal of the Transcultural Nursing Society/Transcultural Nursing Society, 27*(3), 210–217. https://doi.org/10.1177/1043659615592677

Freire, P. (1970). Cultural action for freedom. http://www.thinkingtogether.org/rcream/archive/110/CulturalAction.pdf

Greenwood, M. R. C., & Riordan, D. G. (2001). Civic scientist/civic duty. In *Science Communication*, *23*(1), 28–40. https://doi.org/10.1177/1075547001023001003

Gregory, J., & Miller, S. (1998). *Science in public.* Perseus Books.

Grosfoguel, R. (2007). The epistemic decolonial turn. *Cultural Studies of Science Education, 21*(2–3), 211–223. https://doi.org/10.1080/09502380601162514

Grosfoguel, R. (2008). Para descolonizar os estudos de economia política e os estudos pós-coloniais: Transmodernidade, pensamento de fronteira e colonialidade global. *Revista Crítica de Ciências Sociais, 80*, 115–147. https://doi.org/10.4000/rccs.697

Hacker, K. (2013). *Community-based participatory research*. SAGE Publications. https://play.google.com/store/books/details?id=VTh1AwAAQBAJ

Haywood, B. K., & Besley, J. C. (2014). Education, outreach, and inclusive engagement: Towards integrated indicators of successful program outcomes in participatory science. *Public Understanding of Science, 23*(1), 92–106. https://doi.org/10.1177/0963662513494560

Holliman, R., & Whitelegg, E. (2008). Investigating science communication in the Information Age. In E. Scanlon, S. Smidt, & J. L. Thomas (Eds.), *Communicating science in the Information Age* (Vol. 1). Oxford University Press.

hooks, b. (2014). *Teaching to transgress*. Routledge. https://play.google.com/store/books/details?id=fhIiAwAAQBAJ

Humm, C., & Schrögel, P. (2020). Science for all? Practical recommendations on reaching underserved audiences. *Frontiers in Communication*, 5. https://doi.org/10.3389/fcomm.2020.00042

Hurtado, O. (2007). *Las costumbres de los ecuatorianos*. Editorial Planeta.

Jasanoff, S., Markle, G. E., Petersen, J. C., & Pinch, T. (Eds.). (1995). *Handbook of science and technology studies*. Sage.

Jojola, T., & Shirley, M. (2017). Indigenous planning: Replanting the roots of resistance 1. In S. Kenny, B. McGrath, & R. Phillips (Eds.), *The Routledge handbook of community development* (pp. 324–339). Routledge.

Kling, M. (1963). Violence and politics in Latin America. *The Sociological Review, 11*(1_suppl.), 119–132. https://doi.org/10.1111/j.1467–954X.1963.tb03088.x

Krause, M. (2016). "Western hegemony" in the social sciences: Fields and model systems. *The Sociological Review, 64*(2_suppl.), 194–211. https://doi.org/10.1111/2059–7932.12008

Krauss, L. M. (2015). Scientists as celebrities: Bad for science or good for society? *The Bulletin of the Atomic Scientists, 71*(1), 26–32. https://doi.org/10.1177/0096340214563676

Lane, N. F. (1997). An open letter to scientists and engineers: Let's get the word out together about why science matters. *National Science Foundation.* http://www.nsf.gov/od/lpa/news/media/nlaaultr.htm

Leach, J., Yates, S., & Scanlon, E. (2008). *Models of science communication*. Open University Press.

Ledin, A., Bornmann, L., Gannon, F., & Wallon, G. (2007). A persistent problem. Traditional gender roles hold back female scientists. *EMBO Reports, 8*(11), 982–987. https://doi.org/10.1038/sj.embor.7401109

Lee, Y. O., Curry, L. E., Fiacco, L., Henes, A., Farrelly, M. C., Nonnemaker, J. M., Hoffman, L., & Walker, M. W. (2019). Peer crowd segmentation for targeting public education campaigns: Hip hop youth and tobacco use. *Preventive Medicine Reports, 14*, 100843. https://doi.org/10.1016/j.pmedr.2019.100843

Levitas, R. (2005). *The inclusive society? Social exclusion and new labour.* Springer. https://play.google.com/store/books/details?id=65-JDAAAQBAJ

Lin, A. C. (2019). President Trump's war on regulatory science. *Harvard Environmental Law Review, 43*, 247. https://heinonline.org/hol-cgi-bin/get_pdf.cgi?handle=hein.journals/helr43&section=11

Martin-Barbero, J. (1993). *Communication, culture and hegemony*. SAGE Publications.

Melling, A., & Pilkington, R. (2018). *Paulo Freire and transformative education: Changing lives and transforming communities.* Springer. https://play.google.com/store/books/details?id=gqlVDwAAQBAJ

Merriam-Webster. (n.d.). Latinx. Retrieved May 14, 2022, from https://www.merriam-webster.com/dictionary/Latinx

Mignolo, W. D. (2021). *The politics of decolonial investigations*. Duke University Press. https://doi.org/10.2307/j.ctv1smjncs

Mooney, C. (2017). Historians say the March for Science is "pretty unprecedented." *Washington Post.*

Moreira, C., & Diversi, M. (2011). Missing bodies: Troubling the colonial landscape of American academia. *Text and Performance Quarterly, 31*(3), 229–248. https://doi.org/10.1080/10462937.2011.573190

Mortari, L. (2015). Reflectivity in research practice: An overview of different perspectives. *International Journal of Qualitative Methods, 14*(5), 1609406915618045. https://doi.org/10.1177/1609406915618045

Mutua, K., & Swadener, B. B. (2004). *Decolonizing research in cross-cultural contexts: Critical personal narratives*. SUNY Press. https://play.google.com/store/books/details?id=l3jl-ObBxHQC

Nan, X., & Thompson, T. (2020). Introduction to the special forum on "public health communication in an age of COVID-19." *Health Communication, 35*(14), 1705–1706. https://doi.org/10.1080/10410236.2020.1840754

Nelson, M. P., & Vucetich, J. A. (2009). On advocacy by environmental scientists: what, whether, why, and how. *Conservation Biology: The Journal of the Society for Conservation Biology, 23*(5), 1090–1101. https://doi.org/10.1111/j.1523–1739.2009.01250.x

NIAID. (n.d.). Anthony S. Fauci, M.D., NIAID Director. National Institute of Allergy and Infectious Diseases; National Institute of Allergy and Infectious Diseases. Retrieved May 15, 2022, from https://www.niaid.nih.gov/about/director

Nisbet, M. C., & Scheufele, D. A. (2009). What's next for science communication? Promising directions and lingering distractions. *American Journal of Botany, 96*(10), 1767–1778. https://doi.org/10.3732/ajb.0900041

NSF. (2004). National Science Board: Science and engineering indicators. (Vol. 1). National Science Foundation. Retrieved from: http://webapp1.dlib.indiana.edu/virtual_disk_library/index.cgi/6499571/FID723/seind04/pdf/front.pdf

NSF National Science Board. (2016). Science and engineering indicators. https://files.eric.ed.gov/fulltext/ED571833.pdf

Nussbaum, M. C. (1990). *Love's knowledge: Essays on philosophy and literature*. Oxford University Press. https://play.google.com/store/books/details?id=oq3POR8FhtgC

Padilla, A. (2019). La identidad ecuatoriana por oposición y ofensa. El olvido del mestizo. *Revista de Comunicación y Cultura, 2*, 125–137.
Porter, L. (2010). *Unlearning the colonial cultures of planning*. Ashgate Publishing, Ltd. https://play.google.com/store/books/details?id=rGLKlRHpGcEC
Rainey, G., & Rainey, J. (2017, January). Whistleblowing, the war on science, and the impending Trump administration: Methods of assault and defense in an accelerating conflict. In *Proceedings of the Conference of the Southern Political Science Association*. New Orleans.
Rainie, L., Funk, C., Anderson, M., & Page, D. (2015). *How scientists engage the public*. Pew Research Center.
Reich, C., Price, J., Rubin, E., & Steiner, M. A. (2010). Inclusion, disabilities, and informal science learning. A CAISE Inquiry Group report. Center for Advancement of Informal Science Education. https://www.informalscience.org/sites/default/files/InclusionDisabilitiesandInformalScienceEducation.pdf
Rincón, O. (2016). Los datos: la cancha donde se juega la democracia. *Chasqui: Revista Latinoamericana de Comunicación*, (131), 21–35.
Ross, A. D., Struminger, R., Winking, J., & Wedemeyer-Strombel, K. R. (2018). Science as a public good: Findings from a survey of March for Science participants. *Science Communication, 40*(2), 228–245. https://doi.org/10.1177/1075547018758076
The Royal Society. (2017). The Royal Society. https://royalsociety.org/about-us/history/
Rubin, A. (2021). A look at the complex issues of science communication: The Routledge handbook of public communication of science and technology. *Journal of Science Communication, 20*(5), R02. https://doi.org/10.22323/2.20050702
Smith, L. T. (1999). *Decolonizing methodologies*. Otago University Press.
Stilgoe, J., Lock, S. J., & Wilsdon, J. (2014). Why should we promote public engagement with science? *Public Understanding of Science (Bristol, England), 23*(1), 4–15. https://doi.org/10.1177/0963662513518154
Subbaraman, N. (2021). After Trump, US researchers urge Biden to block political meddling in science. *Nature, 596*(7871), 174–175. https://doi.org/10.1038/d41586-021-02132-2
Swadener, K. M. B. B. (2004). *Decolonizing research in cross-cultural contexts: Critical personal narratives.* SUNY press.
Tervalon, M., & Murray-García, J. (1998). Cultural humility versus cultural competence: a critical distinction in defining physician training outcomes in multicultural education. *Journal of Health Care for the Poor and Underserved, 9*(2), 117–125. https://doi.org/10.1353/hpu.2010.0233
Tuck, E. (2009). Suspending damage: A letter to communities. *Harvard Educational Review, 79*(3), 409–428. https://doi.org/10.17763/haer.79.3.n0016675661t3n15
Van Even, P., & Zaman, B. (2020). Report on needs and wishes for participatory science communication (tools). ParCos Deliverable 4. 1. Report on Needs and Wishes for Participatory Science Communication (Tools). https://lirias.kuleuven.be/3479091?limo=0

Vásquez-Guevara, D. H. (2019). *Science for all: Exploring science communication for public engagement in culturally diverse scenarios in the Americas*. (Publication No. 13812526.) [Doctoral dissertation, University of New Mexico]. ProQuest Dissertations Publishing.

Walsh, C. (2009). Interculturalidad, estado, sociedad. [Interculturality, State and Society]. Abya-Yala.

Walsh, C. (Ed.). (2016). Gritos, Grietas y Siembras de vida entretejeres de lo pedagógico y lo decolonial. *Pedagogías decoloniales prácticas insurgentes de resistir* (p. 17). https://books.google.com/books?hl=en&lr=&id=UONZEAAAQBAJ&oi=fnd&pg=PA17&dq=Walsh+C+(2017)+Gritos+grietas+y+siembras+de+vida+Entretejeres+de+lo+pedag%C3%B3gico+y+lo+decolonial+%5BScreams+cracks+and+sowings+of+life+Intertwining+pedadogy+and+decoloniality+%5D+In+C+Walsh+(Ed+)+Pedagog%C3%ADas+decoloniales+pr%C3%A1cticas+insurgentes+de+resistir+(vol+&ots=K50w781-9Z&sig=n25CnD0VR0zk0_0_QGDu-RR2-gQ

Wynne, B. (1995). Science and the media. In S. Jasanoff, G. E. Markle, J. C. Petersen, & T. Pinch (Eds.), *Handbook of science and technology studies* (pp. 361–389). Sage.

Chapter 3

Methodological Alternatives that Promote Public Engagement for Science Communication

Judith McIntosh White, Angelica Ochoa-Aviles, and Denisse Vásquez-Guevara

Science communication has dramatically evolved in the past two decades from dissemination strategies to more dialogic communication that requires reflecting on how to foster interaction, interest, and participation among non-expert audiences and scientists. For this reason, several scholars have suggested developing science communication through the paradigm of participatory methodologies for research and practical initiatives (Bucchi & Trench, 2014; Haywood & Besley, 2014; Pearce et al., 2015). However, the literature on the subject indicates that developing participatory frameworks for research and practical initiatives has been challenging, and frequently, deficit-based strategies tend to be reproduced even through efforts of dialogic initiatives, such as science communication events and town halls in public museums (Davies, 2008).

Developing research and practical initiatives that promote interchange between scientists and non-experts from the broader public requires learning more about public engagement and participatory methodologies. To overcome deficit-based science communication, participatory action research (PAR) has been suggested by science communication scholars as a promising methodological approach (Bucchi & Trench, 2014; Holliman et al., 2008; Leach et al., 2008) that can guide more nuanced paradigms for research and practice and develops innovative strategies that unite researchers and audiences around science.

Throughout this chapter, we aim to provide graduate students, researchers, and practitioners with several methodological considerations, research procedures, and case studies that illustrate the emerging challenges and guidelines for future studies, and the design of practical initiatives.

PARTICIPATORY APPROACHES FOR SCIENCE COMMUNICATION

Science communication implies not only focusing on the goals of the researchers but rather developing an inclusive and collaborative approach that provides voice and agency to audiences in the process. Eduardo Galeano in his book *Children of the Days* presents a short story that can provide more insights about how non-expert audiences perceive scientists and academic research and what we academics must consider to develop engaging initiatives.

September 7

The visitor

> *In the days of the year 2000, one hundred and eighty-nine countries elaborated the Declaration of the Millennium, which committed them to solve all the dramas of the world. The only objective that has been met was not on the list: it has been possible to multiply the number of experts needed to carry out such difficult tasks. According to what I heard in Santo Domingo, one of those experts was touring the outskirts of the city when he stopped in front of Doña María de las Mercedes Holmes' Chicken Coop and asked her:*
>
> *"If I tell you exactly how many chickens you have, will you give me one?"*
>
> *And he turned on his touch screen tablet, activated the GPS, connected through his 3g cell phone with the satellite photosystem, and started the pixel counter:—You have one hundred and thirty-two chickens. And he caught one.*
>
> *Doña María de las Mercedes did not remain silent:—If I tell you what you work at, will you give me back my chicken? So, I tell you: You are an international expert. I found out because you came without anyone calling you, got into my chicken coop without asking permission, told me something I already knew, and charged me for it.*
>
> —*Eduardo Galeano, Children of The Days* (Galeano, 2013)[1]

This brief passage from Galeano's book (2013) illustrates the cultural collisions between experts or scientists, and non-experts in Latin America. Nonetheless, as researchers who have also worked in other countries through research collaborations, we have also felt identified with this situation when we have met the audiences of our projects. How many times have our scientific lenses been a barrier to engaging meaningfully with the real needs of people in our studies? We have frequently found that our very own studies do not respond to solving the core issues that research participants have and face every day in their lives around health, education, and gender equality among other contemporary concerns. Additionally, as researchers, how times, have we tended to impose research agendas that mostly benefit us and can do little to add valuable resources for our audiences? Moreover, researchers tend to assume that people do not have enough knowledge about their own needs and problems when, in fact, their local knowledge is rich enough to develop feasible and accessible solutions that, combined with scientific data, could yield groundbreaking solutions.

Science communication is a powerful tool to combine research and practice. However, as mentioned before, this discipline can be enriched through the use of participatory action research (PAR) to overcome theoretical and methodological shortcomings (Bucchi & Trench, 2014). Moreover, through our experience as researchers, university professors, and public information officers, we have applied PAR methodologies in studies and initiatives that have brought together researchers and non-experts, guided by empirical evidence of scholars from multidisciplinary studies in health and science communication. Regarding scholarly sources, the current state of the art demonstrates that the participatory approach has been suggested as the future avenue of exploration in science communication research and practice.

In health and public health studies, one might consider that health is a complex process that involves multiple aspects, both at the individual and environmental levels. Health promotion focuses on implementing strategies to prevent risk factors for certain diseases. Understanding the distribution of diseases, their risk factors, and society's perceptions and needs is key to designing effective strategies. Evidence shows that effective strategies are based on behavior change theories and involve the community. In this sense, participatory research is key to designing strategies that are not only effective but also culturally appropriate.

Participatory Action Research (PAR) has been used in numerous health and public health studies (Ball et al., 2021; Clover, 2011; Wallerstein & Duran, 2006), giving it credibility as an effective and inclusive methodology applicable to both research and practice. Additionally, practitioners of science communication for public engagement (Bowater & Yeoman, 2013; Trench, 2021), the newest research movement, suggest that PAR as a methodological paradigm can support the development of more-inclusive research and, in so doing, can overcome many contemporary science communication challenges. Over the last two decades, most inquiry had addressed the introduction of dialogical practices among researchers and their target audiences (Brossard et al., 2005), for example, through citizen science (Cooper, 2016), by focusing on media reporting on science (Dunwoody, 2014), or by exploring the effects of using social media to communicate science as well as such effects on scientific content itself (Brossard & Scheufele, 2013; Lee et al., 2018).

PAR has its origins in the interrelated traditions of action research (Lewin, 1946) and co-learning processes (Freire, 2000). Action research makes use of cyclical problem-solving processes, promoting inclusive participation in planning, analyzing, and implementing different solutions (Minkler, 2004), while co-learning involves a researcher/practitioner facilitating dialogue and capacity building through interaction, interchange, and mutual learning (Freire, 2000). Building on work by Lewin, Freire, and others, PAR studies unite experts and citizens around mutual interests, with the two groups working as co-researchers participating in an egalitarian framework to find solutions to a given topic (Chevalier & Buckles, 2013; Hacker, 2013).

PAR methodologies, then, can bring new opportunities to health/public health research and practice, allowing scholars/practitioners and members of the public to jointly explore a community's health resources and problems and move toward mutually acceptable solutions. Decolonial scholars in the Global North (Porter, 2010; Smith, 2021) have applied the principle of *humanizing planning* by implementing health interventions that involve culturally diverse populations affected by oppressive power structures that relegate them to vulnerable positions. Also, health/public health communication researchers should acknowledge the contextual reality of "scientific rigor" and "objectivity" as they affect people and become aware of the harms inflicted through the very process of their research, to correct for these unintended effects. Consequently, researchers are challenged not only to create knowledge but also—through community collaborations—to identify and suggest acceptable practical solutions leading to tangible actions people can take to overcome health-related problems (Vásquez-Guevara et al., in press). Such work will also comport with the Ottawa Charter for Health Promotion (1986), which lists as prerequisites for health "peace, shelter, education, food, income, a stable ecosystem, sustainable resources, social justice, and equity."

On the other hand, science communication studies, have claimed to include participatory approaches to bring scientists and people together. Earlier studies in the 2000s suggested that participation in practical initiatives could use websites and open access resources about the science regarding environmental studies and health, accessible to people (Nisbet & Scheufele, 2009).

However, during that decade, the transition from the Public Understanding of Science (PUS) movement to the paradigm of Science Communication for Public Engagement (PEST) of the 2010s faced many criticisms. Activities designed by following PUS frameworks tried to incorporate public participation by providing abundant scientific data to the public, but science education initiatives had few opportunities for dialogue and collaboration (Kurath & Gisler, 2009). As participatory practices were incorporated as research final products, another critique was that these practices were designed to benefit the convenience of researchers instead of the public's participation (Horst & Michael, 2011). For example, events merely reported results but were not contextualized to the needs of local communities, and even fewer cases presented opportunities for future collaboration or at least collected ideas from audiences to direct future research.

Literature from the mid-2010s focused more on the paradigm of analyzing the feasibility and benefits that participatory research and practices can bring to science communication. A result of the practices of PUS, studies reflected that science communication needed to start with a bottom-up approach that is inclusive and egalitarian, where researchers and people reunite in dialogue to identify their main interests, needs, and concerns (Haywood & Besley, 2014), and to formulate goals, procedures, and scope of studies and initiatives.

From 2015 to 2020 academic journals and available literature focused on multidisciplinary approaches to develop participatory science communication initiatives. Consequently, newer studies developed in the last five years suggest that technology and traditional media must be combined to ensure participation opportunities in science communication through the use of social media and the benefits of establishing dialogue among researchers and laypeople (Collins et al., 2016; Pearce et al., 2015), arts-based strategies to promote inclusive science communication participation (Dawson, 2018; Takach, 2017), and the intersections of citizen science and science communication as convergent strategies (Bonney, 2021; Constant & Roberts, 2017).

Currently, science communication is exploring participation through PAR and more inclusive research frameworks that foster intergenerational work, and mutual learning. In other words, scientists, science communication scholars, and practitioners are acknowledging that, as The Visitor in Galeano's book, we do not have all the answers and do not need to create complex truth representations through scientific tools and theories that are evident to people. We are currently experiencing a moment of acceptance of all voices and knowledge that matters to build meaningful and useful science communication resources accessible to non-experts of all ages; these resources can also serve people's daily life decision-making processes regarding their health, education, environment, and other fields.

Overcoming Hierarchies and Developing Egalitarian Projects Through PAR Approaches

Still, science communication requires overcoming hierarchical practices and assumptions that perpetuate privileged or dominant groups in different societies as well as in scientific spheres. As presented in chapter 1, there are newer science communication approaches that consider some theoretical models of participatory communication and citizen science. In chapter 2, we offered a broader reflection on building inclusive research and practical initiatives through the guidelines of decolonial thought. In order to propose strategies to overcome hierarchies among researchers and people, this section will explore two streams: first, equality and inclusion, and second, participatory methodologies applied to science communication research and initiatives for public outreach.

The seminal work of Jurgen Habermas (1985) on the development of the public sphere offers important caveats for building inclusive contemporary public spheres around science communication research and health intervention practice. Frequently regarded as providing a prescriptive model of the ideal public sphere, Habermas's original work (Jurguen Habermas, 1985) instead provided scholars with at best a theoretical ideal and, at worst, with merely a description of one public sphere in a single era in time.

Although Habermas (1991) focused on the evolution and decline of the eighteenth-century bourgeois public spheres—elite, bourgeois, and common—had flourished since the dawn of man. However, the history of the common public spheres has been neglected and even eradicated. Thus, there is much work to be done in resurrecting the idea that a public sphere can bring together persons from all rungs of society and may include diversity, mirroring that of the community in which it is formed. Additionally, the public sphere described by Habermas (1991) relied upon face-to-face interaction, letter writing, or print media to convene its members, while today we have at our disposal not only these previous communication channels but also a wide variety of electronically mediated channels, as well as new ways of virtual face-to-face communication.

Today science communication researchers and health intervention practitioners also have at their disposal expanded theories and methodologies to aid in developing participatory and inclusive local, national, and international public spheres around areas of mutual concern, for example, relational dialectics 2.0 and entertainment education methodologies. Thus, the prospect of developing both overarching and targeted public spheres through participatory and inclusive paradigms is bright. It also promises to expand the scope, reach, and efficacy of egalitarian collaborations between researchers, practitioners, and communities as we confront ever more serious issues in our social and physical worlds. It should be remembered, however, that effective and inclusive public spheres should grow organically from the communities where they are intended to function; public spheres cannot be designed by others for communities, nor can they be imposed from outside.

Equality and Inclusion Considerations

Science Communication for STEM initiatives has pioneered the exploration of equality and inclusion practices by developing several analyses and considerations for planning inclusive science education programs. The Center for Advancement of Informal Science Education (CAISE), suggests that inclusion and equality must be placed at the core of STEM initiatives' planning, execution, and evaluation (Bevan et al., 2018). Several of the CAISE considerations acknowledge that scientists from dominant cultures and genders such as White, Western, and Male scientists and audiences have been privileged over minorities in terms of race and gender. To address this issue CAISE suggests that it is necessary to partner with citizen groups, and grassroots community groups to develop discussions about how to put practice inclusion in practice for the needs of their communities around science initiatives. Moreover, partnering with community groups would also offer

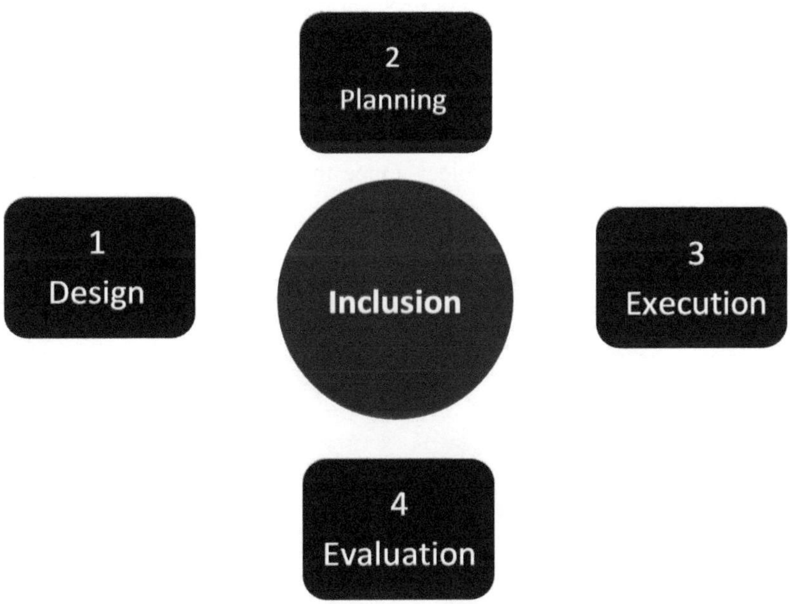

Figure 3.1. Inclusion is a core element of science communication initiatives. ***Developed by the authors from Bevan et al. (2018).***

additional opportunities to engage the audience engagement around science (see figure 3.1).

Science Communication Initiatives as Inter and Transdisciplinary Efforts

Social issues in the contemporary world, such as gender violence, chronic illnesses, global warming, and education disparities are complex and intertwined phenomena that require an inter and transdisciplinary approach to create a better understanding of its roots, causes, and consequences. Moreover, developing science communication projects around these issues is crucial to raise awareness among key stakeholders (i.e., government officials, nonprofit organizations, advocacy citizen groups) in order to foster change and develop useful resources that contribute to the solution of each issue.

Under the design-thinking model to facilitate interdisciplinary, multidisciplinary, and transdisciplinary collaborations to address complex problems, there are three levels of reflection that can be applied to design science communication initiatives (Kalmár & Stenfert, 2020) (see figure 3.2). First, "knowing what" refers to understanding what knowledge is needed to be integrated into a project. Second, "knowing how" is to reflect on which methodologies, interventions, and strategies would be needed to be incorporated. Third, "knowing who" refers to who in terms of the individuals, organizations, stakeholders, and target audiences should be incorporated. However, at this level it is where managing the flow of different timeframes, work styles and schedules of the involved partners or collaborators can be challenging. However, a key tool is to receive training in effective communication skills (Stilgoe et al., 2014), which guides how to communicate transparently with each partner involved in the project. Once the collaboration teams are well

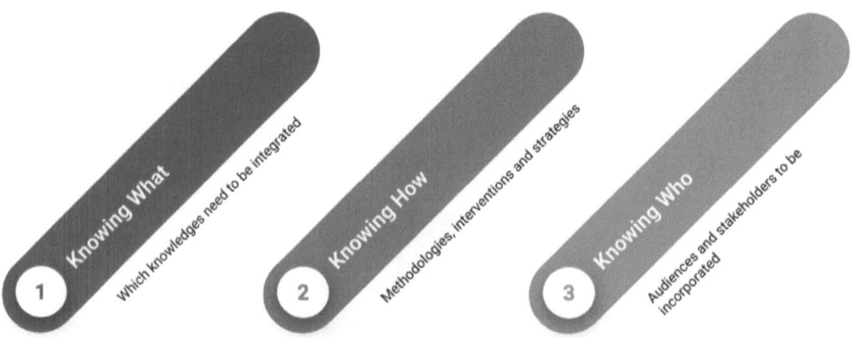

Figure 3.2. Levels for initial reflections in the design-thinking approach. ***Developed by the authors from Kalmár and Stenfert (2020).***

identified, and willing to participate, it is crucial to focus on developing commitments and agreements that are clear and achievable.

Facilitating collaboration in interdisciplinary and multidisciplinary groups is not an easy task. It requires the commitment and trust of the collaborating researchers and practitioners. For science communication collaboration it is crucial to identify specific points of collaboration among researchers or experts of different expertise. For this purpose, useful questions can be the following (Bammer et al., 2020):

1. When is expertise in research integration and implementation required?
2. Where can expertise in research integration and implementation currently be found?
3. What is required to strengthen expertise in research integration and implementation?

These questions can create an outline to identify when each field of expertise will collaborate with others, the teams, resources, and efforts to achieve specific goals. Moreover, after the different collaborators are identified, it is crucial to agree upon the specific collaborations and commitments each expert will take responsibility for and the scope of their contribution. In science communication initiatives, it is important to consider, that collecting data is not only part of the duties of the researchers involved, but also designating a good portion of scientists' contribution to consider the following implications.

1. Scientists involved in science communication must provide useful scientifically supported information—which can be the data produced by their studies or by other authors—and the time to interact with non-experts around responding to doubts or questions of audiences. For example: Informing audiences about the myths around COVID-19 vaccines and the benefits for health prevention in children and youth.
2. Scientists must be open to participating in outreach and engagement events, and communicational resources with non-expert audiences. In science communication projects, a good portion of scientists' time must be devoted to being actively involved in working with communication practitioners and audiences to create useful resources that attend to the of audiences' needs and interests. For example, in a health study attending food safety practices, researchers must be willing to review educational materials such as infographics, and social media posts; moreover, they should be open to providing training sessions for non-experts.

Community-Based Methodologies (CBPR)

A more recent addition to the methodological toolkit of the science communicator and health intervention practitioner is community-based participatory research (CBPR), codified as a method by researchers Nina Wallerstein and Bonnie Duran (2006), among others. Northern health researchers generally credit Wallerstein and Duran as "writing the book" on CBPR, which they summarize as follows:

> Community-based participatory research (CBPR) has emerged in the past decades as an alternative research paradigm, which integrates education and social action to improve health and reduce health disparities. More than a set of research methods, CBPR is an orientation to research that focuses on relationships between academic and community partners, with principles of co-learning, mutual benefit, and long-term commitment, and incorporates community theories, participation, and practices into the research efforts. As CBPR matures, tensions have become recognized that challenge the mutuality of the research relationship, including issues of power, privilege, participation, community consent, racial and/or ethnic discrimination, and the role of research in social change. (p. 312)

Implementation of CBPR-directed projects privileges co-ownership of research and interventions by communities and research/provider organizations. This means that community members are involved from the project's very conceptualization: defining the issues/problems to be addressed by the research project or health intervention; selecting the methods (with input from researchers/practitioners) by which data will be collected or health campaigns conducted; participating in the analysis of data or intervention implementation; making decisions concerning with whom and in what format data will be shared or who will conduct intervention activities; and retaining complete ownership of resulting data or intervention outcomes, including, for example, where data will be stored.

CBPR centers on the process of inquiry into the values, needs, and desires of a community. In this manner, the community members' participation and involvement in the research stages are crucial. In CBPR, researchers and community members serve as co-researchers working together to define the problem, develop a logic model or conceptual frameworks, design the research questions, decide and adapt the methods for data collection, document the interventions, analyze data, and use the information to make an argument (Hacker, 2013).

Conducting science or health communication research or healthcare interventions using the CBPR paradigm is an exercise in cultural humility for Western researchers/practitioners, as well as an exhausting and time-consuming endeavor. This is true so much so that researchers or health practitioners who are on tight institutional or funding timelines are often discouraged from using this methodology. That said, researchers/practitioners who venture down this pathway learn much from the principles on which CBPR rests: "Genuine partnership means co-learning . . . research efforts include capacity building . . . findings and knowledge should benefit all partners . . . CBPR involves long-term commitments to effectively reduce disparities" (Israel et al., 2005, p. 28).

Despite the egalitarian and knowledge-gaining benefits of CBPR, Wallerstein and Duran (2006) outline several paradoxes or pitfalls. The first of these are the problems faced in defining "community participation," which can be seen through a continuum of project control, from complete control by researchers to complete control by the community. They ask, "How do we address the reality that different stakeholders may and do have different goals of participation and different knowledge needs and may and do have different expertise to participate more actively at different stages? Knowledge is never created in a vacuum but rather responds to diverse cultural, social, and material needs of interest groups" (p. 314).

Other issues involve shifting levels of participation away from the CBPR ideal by community members throughout the project; levels of community consent, involving determining the identity of representatives who truly speak for the community and are thus authorized to consent to project activities and processes; "unpacking the role of power and privilege in the research relationship" in order to embrace newer methodologies and theories such as culturally supported interventions based on "indigenous theories of etiology, practices, and programs that emerge from communities" (Wallerstein & Duran, 2006, p. 315); challenges of race, racism, and ethnic discrimination, along with issues of coloniality. However, despite these challenges, evidence supports that CBPR is a potential vehicle for social change and a positive force in reducing health disparities in underserved populations.

RESEARCH DESIGN FOR PUBLIC ENGAGEMENT STUDIES WITH PRACTICAL SCIENCE COMMUNICATION INITIATIVES

Developing science communication initiatives is not an easy task. However, in order to ensure opportunities for public engagement, it is crucial to consider that science communication is still a fairly new discipline, and there are limited grant opportunities to implement and sustain initiatives. For this reason, most science communication initiatives are incorporated as part of research projects as specific deliverables or research products to transfer knowledge to specific stakeholders. Nonetheless, the limited funding and specialized grants for science communication initiatives affect the initiative's scope, duration, and reach have. Even more scarce are the available studies that explore this issue, and the consequence of developing science communication research and the implementation of practical initiatives.

In this section, we will explore and present alternatives to incorporate science communication as part of inter and multidisciplinary projects. We will make several suggestions based on our experiences in writing grants that were successfully funded and that incorporated science communication as components of research projects engaging non-expert audiences who benefited from the results of the study. The effective procurement of research funds involving science communication requires the following considerations (Holbrook, 2010; Lepori & Freco, 2019).

1. Researchers must be aware of calls in their fields. A useful strategy is subscribing to newsletters of national and international funders.
2. Once a potential call is identified, it is essential to read the minimum requirements carefully. Only when it has been ensured that the minimum requirements are met one might proceed with preparing the application. The last ensures that resources and time are optimized.
3. Researchers must guarantee that the application meets the call requirements. Although this seems to be an obvious recommendation, it is vitally important to read the guidelines and ensure that the proposed strategies are perfectly aligned with the expectations of the funding bodies. For example, if a call requires the use of technological platforms, the applicants can propose science communication strategies designed through participatory processes, which will be disseminated through technological platforms suggested by stakeholders and community members.

   Recently, calls for funding requests that proposals support the achievement of the sustainable development goals and include outreach-related activities. Science communication can be perfectly integrated into proposals for such calls. Insufficient communication of science and the lack of use of research results by the community and decision-makers is a widely known problem. The inclusion of a multidisciplinary team and activities aimed at promoting science communication has the potential to translate research results into tangible actions and thus contribute to the Sustainable Development Goals (SDGs).
4. The inclusion of science communication within the research proposal structure must be clear, coherent and concise. In other words, it must be perfectly aligned with the other components of the proposal.

5. Proposals should be prepared carefully and with enough time to ensure a correct understanding of the guidelines and a concise and coherent application. It is helpful to analyze previous proposals that have been funded in previous calls; this will allow for the identification of strategies that have the potential to be welcomed by the evaluators and potentially be more successful. If a proposal is not successfully funded, the feedback and experience gained by the researchers should not be undervalued. It is highly recommended to resubmit an improved proposal in a subsequent call.
6. The research teams must be multidisciplinary and integrate researchers with expertise in the proposed methodologies (e.g., science communication). The methods proposed must be robust and rigorous, and research expertise must be defined clearly. When feasible, several methods must be applied to triangulate the data generated and enhance the result's validity.

How Can Participatory Frameworks Contribute to Developing Science Communication around Research Projects?

Quantitative and qualitative research paradigms may fall short regarding the exploration and facilitation of building collaborations between scientists and non-experts of the broader public. Specifically, qualitative and quantitative studies might be more useful for formative studies to identify needs, interests, and appropriate communication conduits. However, these research paradigms are not so useful in facilitating co-creation, or co-design processes.

For this reason, participatory frameworks can offer more suitable and flexible functions and tools to facilitate collaborative work. For example, in a public health study that has the goal to educate chronic illness patients around self-management strategies that include their treatment and habits, the scientific communication of the study might need to develop a collaboration framework. The actors that should unite in dialogue must be medical specialists, patients, their caregivers, communication practitioners, and researchers to dialogue and explore which are the most suitable strategies to attend to the patient's needs and current knowledge. Moreover, in this dialogue, it will be necessary to carry out an iterative process that leads toward co-creating tangible resources and solutions, such as organizing support groups, a health-monitoring APP, informational campaigns, and outreach initiatives that bring specialists closer to patients.

Participatory frameworks are the most suitable methodology to create a process in which every patient's experience and expert knowledge counts. Additionally, in PAR projects all responsibilities are shared towards reaching a final goal. Moreover, PAR can provide several guidelines to create inclusive research and collaboration settings that can lead to building trust among scientists and people.

The Research Design for Public Engagement Studies with Practical Initiatives

Here we present the results of a health promotion research that applied participatory methods to promote healthy eating and physical activity among adolescents in a city located in the urban area of Ecuador. The study applied the Intervention Mapping (IM) and the Comprehensive and Participatory Planning and Evaluation (CPPE) protocols (Bartholomew et al., 2011; De Vos et al., 2011). Following the IM protocol, firstly, a needs assessment was performed by performing qualitative research with a phenomenological approach (Van Royen et al., 2015; Verstraeten et al., 2016, 2014) and a cross-sectional study to understand the influencing factors on teenagers' eating and physical activity (Andrade et al., 2014; A. Ochoa-Avilés et al., 2012; Angélica Ochoa-Avilés et al., 2014; Verstraeten et al., 2016). The analysis identified some key influencing factors, such as parental permissiveness, role models, peer influence, financial autonomy, skills, and a supportive environment in schools and homes. On the other hand, the CPPE protocol was conducted with the active involvement of the target audiences (i.e., parents, children, and adolescents) to develop a causal model of the teenagers; unhealthy eating, and physical inactivity and to identify possible intervention strategies according to their views.

Based on the needs assessment, eight intervention objectives were defined: adolescents (1) decrease their sugar intake, (2) increase their daily fruit and vegetable intake, (3) decrease their unhealthy snack intake, (4) increase their healthy breakfast intake, (5) decrease daily screen time, and (6) increase physical activity. Additionally, schools offer (7) more healthy food at their food-truck shops and (8) more opportunities for being active. Following the IM protocol, performance objectives were defined for each intervention objective; the performance objectives were contrasted with dietary intake and physical activity behaviors influential factors defined in the needs assessment and in the causal model to generate change objectives for adolescents, parents and school staff. Finally, we identified theoretical methods for each influential factor (i.e., behavioral change techniques) from the literature to define the intervention strategies.

All the processes allowed the co-creation of effective strategies, which were piloted and implemented in a randomized controlled trial with several positive effects on dietary intake and physical activity behaviors (Andrade et al., 2014; Angélica Ochoa-Avilés et al., 2017).

General Considerations for Using PAR Methodologies in Science Communication

As reviewed above, there are multiple ways we can include science communication in grant-writing and research projects to ensure that data findings do not only stay in academic spheres, but that they benefit the community and decision-making non-experts. When research teams are willing to incorporate science communication strategies it is crucial to consider that effective science communication strategies inform people about the benefits, risks, and other costs and consequences of their decisions, and through clear scientific data, they empower people to make well-informed decisions and choices (Fischhoff, 2013).

In this section, we summarize key considerations for using PAR methodologies in science communication projects, specifically for participant recruitment, capacity-building, project design, collaborative data collection and work, data analysis, and research findings presentations.

Institutional Review Boards and Bioethics

As science communication projects might involve working directly with people, it is crucial to be prepared to work with Institutional Review Boards (IRBs) or bioethics committees. These institutional branches frequently are part of research universities, scientific research institutes, hospitals, education systems, and governments. These committees bring together experts in different disciplines of inquiry to make sure research procedures are ethical, non-harmful, and protect participants or co-researchers at all times during projects (Musoba et al., 2015). Getting IRB review and approval frequently requires providing research protocols, descriptions of methods, tools, and overall research goals detailing how processes will be conducted. It is common that protocols require adjustments or changes to comply with the policies and scientific guidelines of each discipline. Once the research procedures are approved by IRB or the bioethics committee, regular reports are necessary to inform about the projects' advancement and their data management.

Recruiting Participants and/or Coresearchers

After obtaining the proper authorizations and reviews before initiating a project, it is crucial to recruit participants or, as PAR projects prefer to call them, co-researchers. Chapter 2 presents several considerations to better understand why science communication could reframe how people are portrayed and involved in projects as actively involved collaborators.

To plan a recruitment process, we suggest that the project team clearly identifies the type of activities, commitments, and compensations participants will have as part of their involvement in the projects. Additionally, like any project, it will be crucial to identify the inclusion and exclusion criteria and add these to the guidelines for recruitment to be openly communicated. For example, these include age range, gender, the topic of interest, type of work, demographic characteristics (education, income, housing), or psychographics (health conditions, life experiences, and interests). Inclusion and exclusion criteria are also useful for screening and selecting participants or co-researchers that match the project's criteria.

After these key points are well identified and placed in the message for participant recruitment, another step is to consider dissemination strategies to reach target audiences. Recruitment strategies frequently use mass media ads on TV, in newspapers, and on bulletin boards at places where target audiences are (i.e., community centers, universities, hospitals), or direct messages through email or SMS through database contacts. Projects must consider ethical and flexible strategies for participant recruitment, for example using social media ads, community centers, associations, or grassroots community groups, and offering them easy access to sign up through websites, online forms, QR codes, and email.

Once an approximate number of ideal participants are recruited, the selection process must be guided by the inclusion criteria. Generally, PAR projects work around groups from five to ten people per group (Chevalier & Buckles, 2019), besides the research team, which is generally one facilitator, two observers to take notes and registry, and one or two researchers.

Once selected the participants or co-researchers will have to provide consent or/and assent forms depending on their age groups while the study is conducted. Several criteria suggest that consent forms must be completed by adult participants after several sessions, but before finishing the project to avoid pressure or biases (Hand et al., 2019). However, in the case of underage groups such as children and teenagers, it is crucial to obtain consent and assent forms from the participants and their legal guardians or parents before the participants initiate their involvement, and they clearly understand what activities will take place and their benefits and compensations (Whittington, 2019).

Capacity Building

Developing participatory projects is not an easy task. It is an effort that requires time, training, the necessary resources, and implementation teams. The role of participants, recruiters, facilitators, observers, and experts in conducting research, or developing an initiative must always consider putting participants or co-researchers first and motivating them to get involved to maximize the benefits and takeaways they can get out of their participation (Minkler, 2004).

For this reason, it is crucial to offer participants or co-researchers the opportunity to obtain new skills, knowledge, and tools through their active involvement in a project. Consequently, in the framework of participatory projects, capacity building is defined as the process through which people gain new experiences, training, and knowledge that later support them in achieving systemic change (Chiu, 2008). Capacity-building can be put into practice by training co-researchers in the use of data collection tools, and software as knowledge transference that is useful for people in their contexts. For example, people can be trained to use mapping techniques to locate specific places that can be resources for the community (Hacker et al., 2012); or they can apply knowledge to real-life scenarios that improve the community's current conditions or benefit it and the researchers involved equally (Wallerstein & Duran, 2010).

Project Design

As we have presented throughout this book, science communication modern frameworks coincide in their new theoretical models and the latest literature that science communication projects need to be developed with the audience's active involvement in order to be effective. As useful guidelines to design a science communication project, we suggest the following considerations.

1. Formative research and secondary studies could be used to inform the project team about the current situation of the target audiences (Lloyd et al., 2012). Data from prior studies or from related areas, along with formative studies could yield key data that would be useful for identifying possible topics of interest, and contextual considerations that are influential from the audience's perspective (i.e., possible focused needs, available resources, ideal communication conduits, possible activities of interest of audiences).
2. In order to work under the criteria of participatory projects, it is crucial to incorporate appropriate compensations for the co-researchers involved in them, as these individuals are bringing their time, effort, and support to collaborate with researchers. Bioethical guidelines for research involving people are critical, according to the principle of maximizing benefits and minimizing harm from the Universal Declaration on Bioethics and Human Rights (UDBHR) and also in *The Belmont Report* (National Commission for the Protection of Human Subjects of Biomedical and Behavioral Research, 1978). However, research should not only about harming participants, but it should also recognize the

labor that projects demand from them. Consequently, a practical way to compensate participants is by providing them monetary payments (Shekhani et al., 2021), which should not fall into large payments that could condition research participants or co-researchers to fall into coercion practices, but instead recognizes the time and effort of the participant in the project (Sofaer et al., 2007). As researchers receive a salary for their work, it is crucial to offer similar compensation to people as well.

3. Grassroots organizations, NGOs, clubs, and even online communities can be a useful resource to locate target audiences and identify key stakeholders (Humm & Schrögel, 2020). We suggest contacting the groups that bring together the target audiences before initiating project planning and invite them to discuss the specific topic of interest around a science communication project. This first approach can be useful in indicating audiences' interests, needs, and ideas to become involved or to determine crucial factors that are important to them. When researchers or communication practitioners first start reaching out to target audiences, they are opening the opportunity to get feedback from direct sources that can enrich research, but also support people's learning about how to solve or address issues that affect them.
4. Once participants or co-researchers are recruited and invited to work on a project, science communication teams must consider initiating dialogue about defining the project's scope and goals, and how collaborations among scientists and participants will happen. Additionally, we suggest focusing the conversation on developing feasible solutions that are applicable to audiences' needs and concerns, for example: offering solutions for food entrepreneurs to comply with food safety training, such as developing educational courses and tools. This audience-centered approach is consistent with participatory methodologies and decolonial guidelines to demonstrate truthful commitment and willingness to transform scientific knowledge into useful and practical alternatives adapted to people's needs (Atkinson & Mason, 2014; Marschalek & Schrammel, 2017).
5. When the science communication team is preparing to start the project, and participants or co-researchers are assigned specific tasks and responsibilities, it is crucial to organize work sessions by thinking about criteria of inclusiveness and accessibility (Massarani & Merzagora, 2014). In practice, choosing convenient locations for work sessions, as well as dates and times, and offering enough access to work materials are different strategies to be mindful of offering a welcoming, collaborative space.

Collaborative Work

One of the challenges around science communication projects is to coordinate collaborative work with groups of different cultures, age groups, and educational levels. Specifically, PAR projects of science communication unite researchers, non-experts of different ages and backgrounds, and communication practitioners in collaborative work. Under the participatory paradigm, projects developed with the assumption that researchers are facilitators of research processes, but the active involvement of participants or co-researchers is crucial with shared agency and responsibility regarding the completion of the tasks involved. In science communication projects, collaboration is defined as a crucial factor that influences audience engagement in science (Riedlinger et al., 2019). Additionally, it is understood as the involvement and interaction between people in coordinated efforts to analyze a problem and builds on collective knowledge that brings alternatives and solutions (Campos, 2022; Fox, 1987).

However, the scope and reach to develop collaborative projects depends on the type of supportive science and technology policies countries, institutions, and grant agencies offer for research. For example, Colombia's policies for science and technology have determined that 10 percent of the country's total royalties from multinational mining companies operating in Colombia will be allocated for scientific research. Moreover, Colombia's national policy demands that scientists must have mandatory social appropriation strategies for the findings of each project (Ayure & Triana, 2022). Consequently, under similar circumstances, researchers must have more resources and valid arguments to put effort into developing projects that are focused not only on knowledge transfer but also on interactions and dialogue with people about what else is necessary to develop inquiry in the future.

Anne Leitch (2022) recently systematized some guidelines for participatory science communication based on how to focus on dialogue and active participation. The author presents the criteria of *place, power, pain,* and *poisson* and the need to challenge some assumptions on participatory science communication. *Power* describes the people who decide who is invited to participate and who will be excluded, and how the collaboration process will unfold. Additionally, under the idea of power is the expertise or knowledge that is useful to the project and needs to be incorporated. *Place* refers to understanding the context and connections between local knowledge and the way collective learning happens in the context of people's culture. *Pain* refers to the difficulties that people have when embracing change in a process of transformation. Moreover, the pain refers to the projected loss and historical roots that might be connected to the resistance to change (i.e., colonization in indigenous groups). Finally, *Poisson* refers to the Lao Tzu proverb "give a man a fish and you feed him for a day. Teach him how to fish and you feed him for a lifetime." Leitch refers to this in the terms of science communication and suggests not handling audiences scholarship that is not well explained and contextualized, but critically reflect on how audiences—from their perspective—want to approach the topic, how necessary it is for them, and how they can use knowledge to transform their lives and social issues (Leitch, 2022).

Consequently, to promote collaboration, dialogue must be the center, tool, and strategy to plan, shape, and execute science communication. Here, the scientists' work gives equal importance to both. First, the effort to build tools and resources that in the process also produce opportunities for mutual learning, trust, and tools that provide useful solutions and new knowledge. And second, to interact with non-expert audiences to find common ground for developing initiatives together.

Co-design

Collaboratory projects in science communication must consider space and processes that enable researchers and participants/co-researchers to co-design initiatives that present scientific content in a way that motivates engagement and presents useful resources for audiences. As an innovative process, co-design is understood as a process in which people from different backgrounds and organizations combine their knowledge and experiences through participatory methodologies with the goals of improvement, evaluation, decision-making, and promoting cooperation (Steen, 2013).

Frequently, science communication is developed with problem-solving goals that specific populations might be facing (Orthia et al., 2021), in order to present data or propose alternative solutions. After we have recruited, trained, and organized work tasks among researchers and participant/co-researcher, it is crucial to focus on identifying which are the key points that can guide a co-design process that leads to an ultimate strategy that creates a useful resource for people and creates new knowledge about science communication.

For example, in Case Study 3 about food safety issues of informal food entrepreneurs, we identified that most of them mainly started their businesses because of unemployment issues as a result of the COVID-19 pandemic. These vendors did not have any formal food safety training, which leads to consumer health problems due to crossed-contamination malpractices and inadequate food storage in domestic kitchens. Consequently, the project that brought together informal food entrepreneurs, food scientists, and science communication specialists agreed to focus, first, on developing educational tools accessible through social media resources and, second on developing training programs that are certified by local universities. Consequently, the project team aimed to develop informational tools, tailored messaging tips for social media, educational manuals, and curricula for training programs that could be offered to entrepreneurs for free or at a low cost.

A co-design process for science communication is required to explore collectively—with researchers and non-experts—the main issues, their causes, and consequences around a specific topic to identify feasible solutions and, from this point on, to reflect on how different strategies can be used to solve specific issues. Thus, we suggest that science communication teams use the action-reflection cycles proposed by McNiff and Whitehead (2009) as an iterative process, which allows them for in-depth discussions that help to explore and find the real needs that should be addressed through science communication.

Action-reflection cycles have four broad steps that drive one another into a cycle as seen in figure 3.3: *plan*, *act*, *observe,* and *reflect* (McNiff & Whitehead, 2009). The step *Plan* refers to identifying what can be done. This process implies exploring the issue, its possible causes, existing or possible consequences, and its emergent implications. Planning also implies

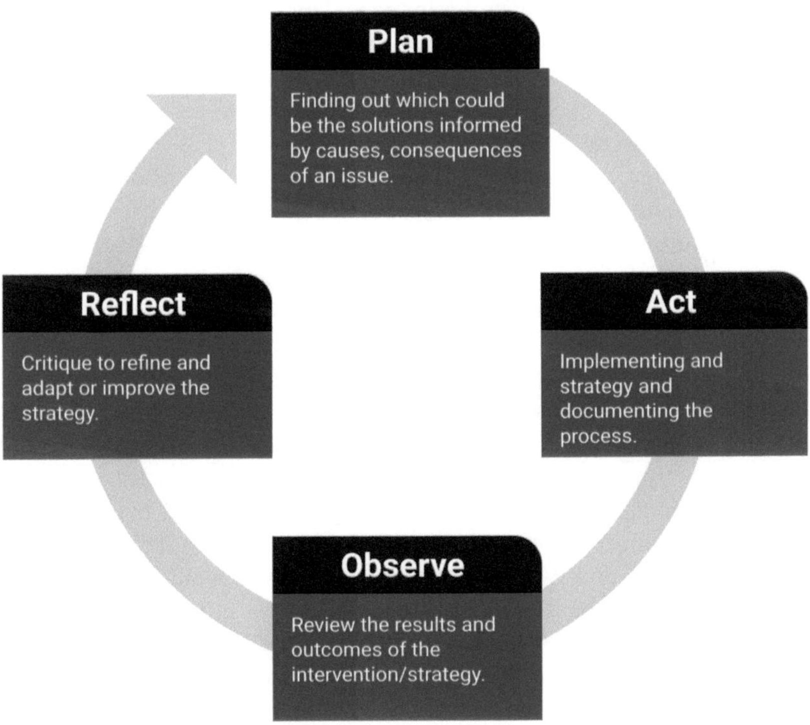

Figure 3.3. Action-reflection cycle process. ***Developed by the authors from McNiff and Whitehead (2009).***

shaping the solutions and how these can be implemented therefore this step involves considering the messages, conduits, operation, and timeline of necessary activities as well as the individuals who will participate or intervene as facilitators or audiences. The following step, *Act,* implies executing the activities planned in the first step. Before a large-scale execution, we suggest pre-testing activities in order to correct any possible errors in the science communication initiative. In this second step, it is necessary to collect evidence to monitor the activities' outcomes, which will be used in the next step of the action-reflection cycle (McNiff, 2016). The third step, *Observe* contemplates the evaluation and review of the outcomes and results of the action step. Moreover, observing can provide additional insights to shape, or modify the initiatives, or add other strategies or resources that could strengthen communication goals. Finally, *Reflect,* refers to developing critiques or questions informed by the evidence of the action step and the observation step, which can lead to developing and improving the science communication strategy.

Participatory Research Design for Public Engagement

Previously, we mentioned several prior considerations to integrate multidisciplinary and collaborative projects where science communication is articulated with the work of researchers or experts. These considerations can be helpful to map resources and expert profiles of experts that are necessary to execute and develop successful projects.

After the project's structure is defined, developing a participatory methodological design according to the scope and goals of the research topic is required. Before explaining the procedures to plan a participatory study, we explore some of the genres of participatory methodologies, constructs, and functions. These paradigms offer different nuances and tools that adapt to participatory work with different audiences, research goals, and contexts.

PAR traces back its foundations to two different but interrelated disciplines, Kurt Lewin's action-research (1946) and Paulo Freire's pedagogical co-learning approach in his book *The Pedagogy of the Oppressed* (Freire, 2000). Lewin, from a psychological approach, referred to action research as a cyclical process to solve problems through people's full participation in the steps of planning, action, and research (Minkler, 2004). This contribution of Lewin to CBPR and PAR is better known as the "emancipator research" (Hacker, 2013) because it empowers people to take action about the issues that affect them in order to solve them. On the other hand, Freire (2000) a Latin American educator, developed another foundation of participatory research methods, through his concept of co-learning. Co-learning proposes a role for the researcher as a facilitator of knowledge production but also argues that research can emerge not only from academia but also from the interests of society working in a community framework, where all the participants are co-researchers and contribute with their knowledge (Hacker, 2013).

PAR, as a methodological approach, fosters the co-construction of knowledge, and an iterative process that circulates developing solutions, implementing them, and later evaluating their outcomes and adapting or improving its results.

Among the main functions and scope that participatory action research (PAR) can offer as a methodological approach (Lake & Wendland, 2018) we find the following opportunities:

1. Aligning the outcomes and resources produced with research to the needs of specific social groups.
2. Moving towards collaborative knowledge production that is applicable and useful to society.
3. Increasing non-expert capacities by teaching new skills and the use of tools for knowledge production.

In terms of working with stakeholders (i.e., government agencies, grassroots groups, NGOs, or any organizations or institutions around a specific topic) and engaging them around any research (Eelderink et al., 2020), PAR studies offer several strategies and open space for the following scenarios:

1. Understanding the system in which the specific problem is embedded, rather than solely the problem's direct and individual context.
2. Understanding the problem context from the stakeholders' perspective to motivate social learning, a cooperative mindset, and innovation power.
3. Creating and implementing a strategic plan based on stakeholders' motivation, needs, and concerns which tackles the problem by improving the system in which the problem is embedded.

PAR also matches decolonial and indigenous epistemologies, as these discipline streams foster empowering audiences in research processes and knowledge co-creation. Consequently, it is recommended to use theories from these streams as epistemological foundations to develop self-reflexivity, ethical procedures, and egalitarian research settings.

For science communication, PAR methodologies can open several opportunities for audience engagement. In the current times of rapid changes and complex social problems, people are becoming aware of the importance of getting involved around topics of public interest, but especially, of learning more about what they can do to contribute to solutions through their daily practices or by putting their efforts in citizen groups or grassroots organizations that pursue social change or advocacy (Hecker et al., 2018). As revised in chapter 1, among the new dialogic models, citizen science is a fairly new model that unites scientists and people around topics of mutual interest for collaborative work. Science communication can also support initiatives by focusing on creating useful informational resources and connecting scientists and people. Nonetheless, for inquiry and operational purposes, the PAR methodology constitutes a powerful tool.

PAR methods can be useful to identify the specific needs of audiences in a tailored fashion, where people can explain what they need and the specific ways they need to learn more. For example, a circular economy inquiry project might be interested in disseminating key practices for recycling, On the other hand, people might be interested in learning more about how to recycle at home properly. PAR methods and tools can help to explore people's current knowledge around recycling, but it can also be useful to co-design routines to teach families how to recycle efficiently at home, how to choose different purchases, and even how to repurpose materials into other useful objects for households.

PAR methodologies offer researchers, whose core commitment is to bring science closer to people many tools to communicate, collaborate and co-create, implement and evaluate current initiatives and how these could be improved, but also create new initiatives tailored to the audiences' needs and interests (Lake & Wendland, 2018).

Once we have explored the main constructs and concepts of the PAR methodological approach, it is necessary to identify which genre of this research paradigm is more suitable for the type of science communication project we will be developing, its goals, scope, and available funding and resources.

PAR Genres

Among some of the participatory research frameworks, we find the following tools:

1. *Action Learning Systems*. Participatory research can be very productive for the organization, community, and multi-stakeholder processes. Frequently, different processes fail to attend to recurrent issues that remain unsolved, but they also represent opportunities for research and to take collective action into collective learning and change (Chevalier & Buckles, 2013). In this section, we explore how action learning systems as PAR approaches work as participatory research genres. Specifically, we explore how these genres can be useful for science communication projects that require fostering collective learning

through the *Action Research and Training* (ART) approach and organizing into action through *Planning, Inquiry, and Evaluation* (PIE).

2. *Action, Research, Training (ART)*. This participatory approach's goal "to assess the current and desired balance and integration of three components within a learning system" (Chevalier & Buckles, 2013). For science communication initiatives the PAR-ART method is very useful to build from zero educational strategies in a collaborative fashion by bringing together experts and the target audiences of the educational strategy. Later on we present a case study that used ART as a participatory method to develop an educational tool for non-experts.

The ART approach includes the following components:

1. Actions, including deciding, planning, and putting actions into to practice to achieve concrete goals.
2. Research through data collection and analysis.
3. Training or teaching and the required capacity building for research participants and co-researchers.

To conduct participatory ART studies, we need to comprehend the following steps:

1. *Defining the focus of the project*. This step's purpose is to define a key project, program, or institution and list major actions, research, and/or training activities undertaken previously, ideally with the last six months as a minimum, up to a year maximum.
2. *Assessment of learning ART components*. This second step assesses and compares the weight and importance of action, research, and training collaboratively among participants or co-researchers and researchers. Venn diagrams are frequently used to find any possible overlaps among components and to promote further discussions to shape the actions that need to be developed.
3. *Analysis of ART components interaction*. If the project prioritizes multiple weights of components, or if there is significant overlapping, for example between research and training, it is necessary to develop actions that attend to both needs.
4. *Review of ART*. Once all the components or action, research, and training are identified it is imperative to compare the current condition to the ideal conditions defined in the collaboration between co-researchers/participants and researchers, to define which actions and strategies

need to be implemented to achieve the ideal state or goals determined in the study.

Case Study 3: Developing an Educational Tool for Ecuadorian Food Entrepreneurs around Food Safety Practices in the Context of COVID-19

In the context of the sanitary emergency due to the COVID-19 global pandemic, the Department of Biosciences, University of Cuenca, the Biotechnological Research Center of Ecuador (CIBE), Escuela Superior Politécnica del Litoral (ESPOL), and the Department of Food Science and Biotechnology (DECAB), Escuela Politécnica Nacional (EPN) led a research project with the purpose of strengthening food safety practices of street food vendors and food entrepreneurs. The funding for the overall study was provided by the Ecuadorian Corporation for the Development of Research and Academia, *Corporación Ecuatoriana para el Desarrollo de la Investigación y la Academia*, Ecuadorian Corporation for the Development of Research and Academia (CEDIA), and the scientific team was supported by the Food Safety Consortia INOAL and VLIR-Network Ecuador. The project, which was originally designed to conduct in-person research in its science communication section had to be re-organized into an online-based study. As street food vendors and food trucks were not allowed to continue at their usual physical locations in cities due to the mandatory confinement imposed as a measure to slow the spread of virus, these small businesses sought new alternatives to keep operating from their homes and opted to promote their services through social media (Facebook and Instagram mainly); they also adopted online payment platforms, and provided home-delivery service for their customers through motorcycles or bikes. Consequently, food preparation was not supervised by sanitary authorities. Moreover, emergent entrepreneurs, who appeared massively due to unemployment, did not have the required sanitary authorizations to be vendors, had not gone through any food safety training, and consequently lacked food safety practices.

The project included food science experts, public health researchers, health communication practitioners, street food vendors, and emergent entrepreneurs in order to (1) analyze through microbiological techniques the quality of foods sold by the entrepreneurs to determine any possible food contamination, and (2) to determine the best educational strategy to develop training about food safety practices. The communications team developed a participatory study using the ART approach mentioned previously above. The main goal of the science communication section was to explore and co-design appropriate teaching strategies that focus on specific key practices to ensure food safety such as avoiding crossed-contamination, appropriate food disinfection, management, and storage, and food handlers' hygiene, and packaging.

The study recruited thirty food entrepreneurs from Quito, Guayaquil, and Cuenca, the three largest cities in Ecuador, through each university's research department's social media accounts (Facebook and Instagram). All the research procedures were revised and approved by the University of Cuenca IRB, and participants were selected through inclusion and exclusion criteria regarding the type of food business they had. As inclusion criteria, the study accepted all entrepreneurs of food trucks, online vending strategies through social media, and access to mobile devices and internet access to participate in the study could participate. Regarding exclusion criteria, food vendors with a venue or belonging to franchises were excluded, as well as entrepreneurs that did not have access to the Internet. Later, selected participants were required to provide a consent form, and they received a $25.00 gift card as compensation for their participation.

The ART-PAR approach enabled the research team to explore food entrepreneurs' and handlers' current knowledge and needs; additionally, made it possible to determine which food safety practices were unclear and needed to be revisited as well as included in a training strategy to educate people about how to avoid food-related illnesses. To conduct the study, we organized four participatory workshops in the three cities through the Zoom video conference platform, each session lasting approximately one hour (see figure 3.4). Each workshop had specific goals that were shared and discussed in the first workshops with the participants in order to provide a flexible research design that also included their needs and concerns.

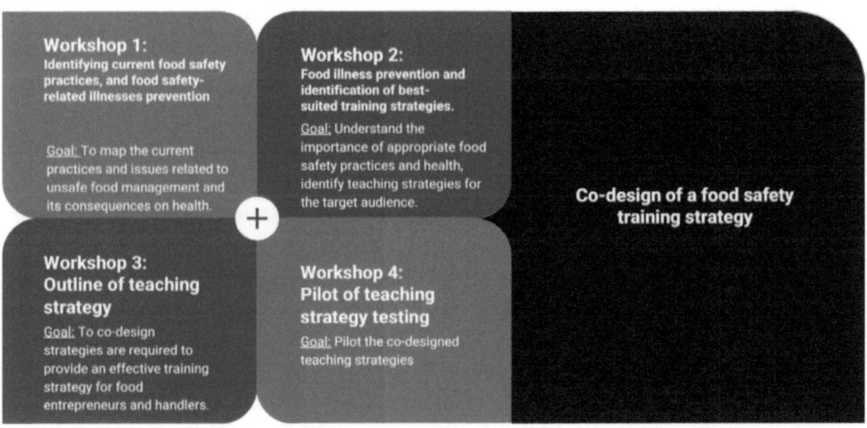

Figure 3.4. Goals of ART participatory sessions for co-designing a training strategy around food safety. ***Developed by the authors.***

The workshops were conducted by a research team of two facilitators, one observer, and one food safety expert. To develop a participatory dialogue, semi-structured prompt questions were developed beforehand and discussed with the participant co-researchers. As the workshops were conducted through *Zoom* given the mandatory confinement of 2020, we used screen-shared digital diagrams adapted through *Google Slides* to collect and present the emergent data which served to visualize issues, their causes, and practices to solve incorrect food safety conditions.

Participants also presented their ideas to develop an effective training approach and tools that would be useful to their businesses. After analyzing the needs, interests, and current knowledge of food entrepreneurs, the overall training strategy co-designed through the collaborative ART approach combined the following components (Chevalier & Buckles, 2013):

1. *Food safety training workshops*. Food scientists at Ecuadorian universities must create an educational training program about food safety training that provides a certificate to entrepreneurs. It was suggested that the program's modules have a direct approach to evaluating the practical skills and knowledge of food entrepreneurs.
2. *Food safety toolkit*. Along with training, food entrepreneurs need to have useful tools to remind them of the daily practices around food safety, such as food disinfection, storage, and packaging, among others. For this purpose, we created a toolkit (see figure 3.5) that included (a) a food safety production chain process adaptable to kitchens and workspaces; (b) cutting boards for each food group (vegetables and fruits, meats, seafood); or (c) a food safety handbook with key practices for food safety.
3. *An online social media and web-based educational community*. As a strategy to strengthen the food safety skills and practices learned through training, each city's team proposed the creation of a community of food entrepreneurs supported by food scientists. This community would be a Facebook fan page (see figure 3.6) where the group could post about tips and tricks to maintain food safety practices, training opportunities, and a space to motivate interactions and interchange among food vendors. Moreover, we developed a landing website to develop video tutorials to instruct about food-related illnesses and other practices.

Social Change PAR Methodologies

Interventions or programs that are oriented towards contributing to solving social problems such as safety, public health, and environmental preservation,

Figure 3.5. Food safety toolkit pilot for food safety training co-designed by researchers and food entrepreneurs. ***Toolkit developed by the Health, Nutrition and Physical Activity Research Group at the University of Cuenca, Biosciences department. Photo courtesy of the University of Cuenca, Biosciences department.***

among other issues, frequently require evaluating their results and outcomes with their target populations. Specifically, in science communication interventions and programs, it is fundamental to assess, the audiences' perspective, how useful and effective strategies were. Moreover, it is crucial to

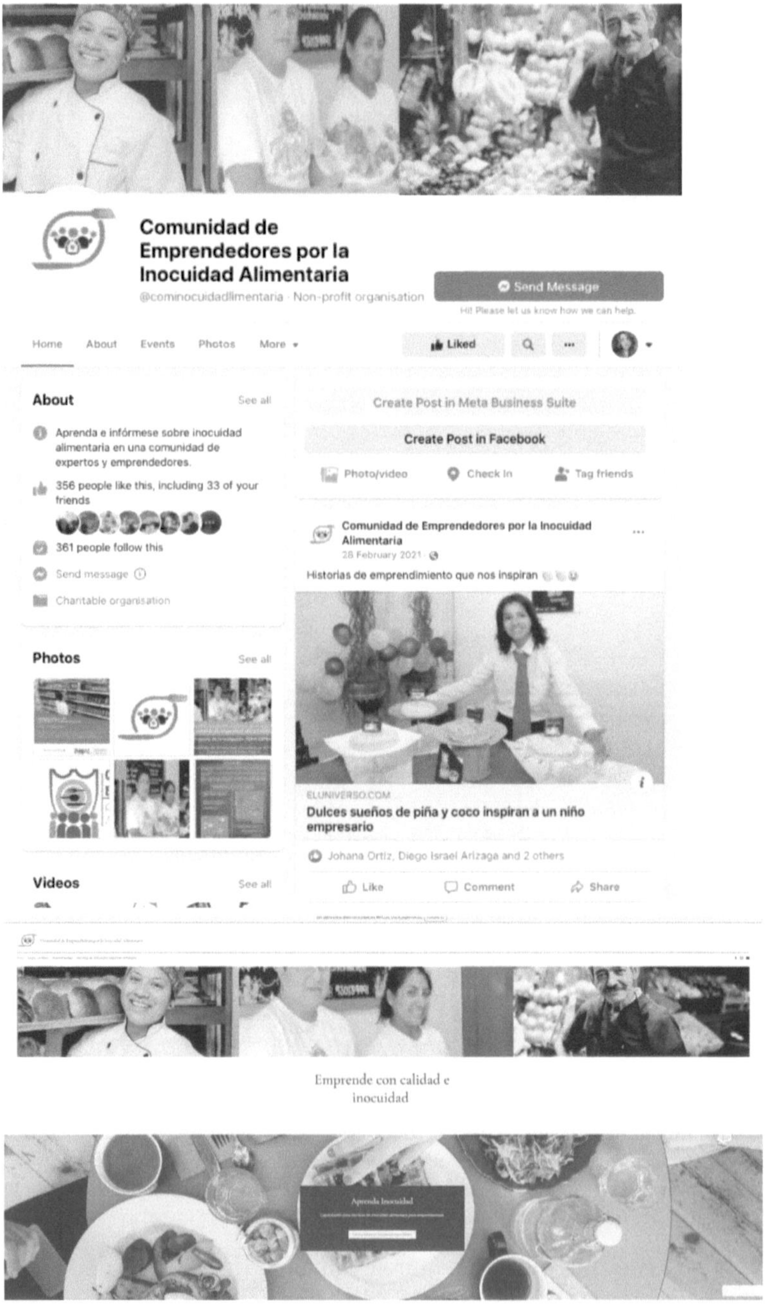

Figure 3.6. Social media and web-based pilots were developed to strengthen food safety training. ***Developed by Health, Nutrition and Physical Activity Research Group at the University of Cuenca, Biosciences department.***

understand that there is always room for improvement and to create newer strategies in a participatory manner. For this purpose, this section will explore PAR methods to develop evaluation studies that can be useful to retool and improve science communication initiatives collaboratively.

Attribution and Contribution Participatory Studies. This methodology helps to identify the findings around the target audiences' suggestions and recommendations from their perspective regarding how an intervention has or has not contributed to their needs and interests (Chevalier & Buckles, 2013). For example, how much change has happened after specific strategies were implemented, which actors should be included, and which are the key influence factors for intervention success. Some of the functions of *Attribution and Contribution Studies* are the following (Chevalier & Buckles, 2019):

- The change that was observed in the audiences.
- Audience understanding of the scope of the intervention.
- Roles of actors and factors and barriers for interventions.
- An evaluation of the methodological performance of an intervention.

The following steps address the attribution and contribution process in PAR studies to access the interventions' or programs' performance. Moreover, in any PAR study, it is necessary to understand the active involvement and collaboration of the beneficiaries or the recipient target audiences:

1. *Intervention Description.* As a starting point, a description of the action, intervention, or project that will be evaluated is necessary. Specifically, the main goals of the intervention, timeframes, and the role of implementing partners need to be clearly identified.
2. *Discussion of Changes after the Intervention.* As a second step, the evaluation participants, the changes observed, and their level of contribution (major, small, or moderate) as positive progress need to be discussed as also any possible negative situations that have not been solved through the intervention strategies. To evaluate the intervention strategies' level of contribution a diagram can be used to rank the contribution levels of each of the strategies that are part of the intervention, according to the audience's discussion. Chevalier and Buckles (2013, 2019) suggest using a scaled-level diagram to rank and evaluate the contributions as seen in figure 3.7.
3. *Scope of Intervention.* In a third section, this point discusses the role of interventions as sole contributions or if they play a joint role in other parallel interventions. For example, in the case of health prevention campaigns for COVID-19, if the governments' Ministries of Public Health had major guidelines or policies that raised awareness about

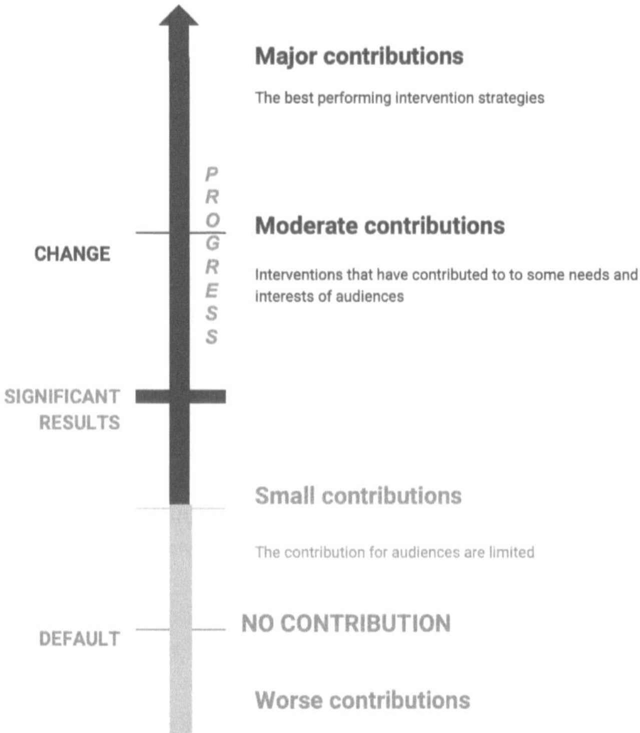

Figure 3.7. Attribution and contribution scale diagram for representing the level of contributions of intervention strategies. *Developed by the authors from Chevalier and Buckles (2013, 2019).*

prevention that matches the similar messages of other prevention campaigns.

Moreover, the discussion must focus on describing how the specific intervention strategies contributed to driving changes in the target audience. This section of the study also must explore any barriers or obstacles audiences faced in to accessing or putting into practice the suggested actions or behaviors.

In terms of communication, a discussion about how effective the conduits to disseminate messages were effective to each audience needs to take place. Additionally, the usefulness and attractiveness of the message's content must be discussed to suggest any changes or adaptations that are more engaging and interesting for audiences.

The attribution-contribution methodological framework can be very useful to science communication initiatives for evaluation studies that invite audience participation. Frequently, these initiatives have been assessed through quantitative and qualitative studies, nonetheless, these research approaches have the major limitation of addressing audiences only as consultation studies. For example, museums frequently evaluate ticket buyers by using surveys in-situ through electronic devices or emails. However, existent evidence confirms that these evaluation methods are simplistic, and they might not reflect in detail variables such as scientific learning and audience opinions (Jensen, 2014).

To evaluate and improve existing science communication initiatives, in-depth explorations—through qualitative and participatory research—have been evidenced to be more effective, as they provide audiences a space to co-analyze them and directly collaborate by suggesting changes or adaptations to specific strategies or communication tactics used (Davies, 2019; Enros & Bandelli, 2018). There are multiple participatory approaches to conduct workshop sessions and collaborative research that are useful and can be adapted to the specific needs of science communication initiatives. However, it is crucial to understand that the active participation and presence of the scientists who authored and are experts in the field is essential to clarifying information that emerges from non-expert audiences but also to building partnerships and dialogue in the process.

Frequently, evaluation of science communication initiatives is delegated to consultants and communication researchers who are not experts in public health, gender studies, or any specific scientific discipline the initiative addresses. Not having discipline experts as part of the participatory evaluations can also lead to their failure by developing new strategies in which information is not well supported by scientific data. For this purpose, experts also need to participate in an egalitarian manner with their audiences, by actively listening to their suggestions and doubts, and also by learning more about the needs, concerns, or interests they have. This effort is key to build trust with audiences, and further on, to develop partnerships that can lead to new research, science communication initiatives, and other positive opportunities for public engagement and outreach.

Participatory Rural Appraisal (PRA)

As a participatory research genre, participatory rural appraisal (PRA) enables local people to share, enhance and analyze their knowledge of life through their cultural identity practices (Chambers, 1994). As a methodology, PRA focuses on "building upon the strengths that are already present in communities" (Chandra, 2010). It is considered one of the most popular and effective participatory methodological approaches to gather information and develop collaborative studies through a bottom-up approach where communities share their experiences and explore how to create solutions by using their local knowledge (Cavestro, 2003).

For international development work and research, PRA and its later successor Participatory Learning Action (PLA), rapid rural appraisal (RRA), and PRA are the most known and used participatory approaches (Chevalier & Buckles, 2013). PRA and RRA broaden the idea of rural development to explore the broad aspects implied in the livelihood and use of natural resources and local knowledge to find sustainable solutions for rural communities.

Some principles for PRA are the following (Cavestro, 2003):

1. *Participation.* The center of the method is to provide community participation. Consequently, in PRA data collection is not only, but it also promotes people's involvement in the process as research or project partners that also have the shared responsibility—along with researchers—to gather and analyze information.
2. *Flexibility.* The work or research teams must be trained to apply flexible data collection tools and facilitation skills adaptable to the size and mix of the PRA participants team, their needs, and capacity.
3. *Teamwork.* The PRA facilitation teams must be preferably local teams that speak the community's local language and know its current needs and reality. Moreover, the PRA facilitators must be diverse specialists (i.e., social scientists, health sciences) and procure gender equality for women and men representation.
4. *Optimal Ignorance.* In terms of efficiency, PRA tries to be efficient in terms of both time and money. PRA work has the goal to collect enough information to make the necessary recommendations and decisions.
5. *Systematic.* PRA data is a useful approach to develop formative research through a qualitative approach in small sample sizes. However, alternatives to conducting larger PRA studies that ensure validity and reliability of findings are for example including sampling based on approximate community stratification by geographical location, and cross-checking data through already existing statistical studies.

Rapid Rural Appraisal (RRA)

This participatory approach focuses on developing research in short periods, by quickly proceeding to collect data, analyze it and evaluate the conditions of the local population in rural areas (Cavestro, 2003). To develop RRA, interdisciplinary teams are required due to the complex exploration of issues and needs that rural people might have. To conduct successful RRA studies, researchers must consider having a well-developed data collection toolbox, which enables people with several inequalities to participate (i.e., illiteracy, communication needs, or physical and mental limitations). Consequently, it is suggested that researchers use data collection tools focused on dialogue and visual representation such as mapping, ranking, and participatory diagramming, by using symbols, drawings, and prompt questions to guide discussions.

RRA has been used largely to develop work as a form of interdisciplinary and participatory evaluation research (Herr & Anderson, 2014), and in doctoral dissertations that focus on participatory work with intergenerational groups with cultural diversity and different education levels.

Among the participatory genres, RRA and PRA have demonstrated to be useful methods to undertake science communication research that aims to work with local communities around education and to discuss solutions regarding the prevention and awareness about climate change or natural preservation in collaborative projects. Researchers of the University of Kisumu in Kenya used PRA to explore and codesign educational tools about the impact of climate change in the area of Mara River Basin, which is a zone that is deeply affected by the drainage of basins in Lake Victoria. For this study, researchers were required to explore how to develop climate change knowledge and education. For this purpose, they brought together twenty-four community organizers (men, women, and elders) in participatory workshops. Mapping and participatory diagramming techniques were used to discuss how the local landscape of their community has changed, and how to identify practices that can harm the environment and organizations that are key partners to prevent impact (Omondi, 2020).

For RRA and PRA, working with people's tools, knowledge, and local resources of people for data collection, analysis, and co-design strategies is crucial. For example, using symbols, objects, notes, or even handmade tools such as drawings, paper cuts, and posters and applying local knowledge that people have to create feasible initiatives that become useful and attractive to them to learn more about science.

Youth Participatory Action Research (YPAR)

As a methodology, YPAR provides young people not only the opportunity to participate in studies of social problems that affect their lives at different levels but it also invites them to propose solutions to address these social problems by using their social capital and local knowledge (Cammarota, 2017). As a methodological approach of the PAR paradigm, YPAR focuses on groups of young audiences—teenagers or young adults—to develop collaborative studies where they act as co-researchers along with researchers from different disciplines.

Theoretical streams frequently used to support YPAR are critical race theory, intercultural studies, and communication, gender theories, decolonial thought, urban justice studies, and pedagogical emancipatory theories (Fernández-Aballí Altamirano, 2020; Fine, 2009; Tuck & Yang, 2013).

As a methodological genre of PAR, YPAR projects embrace a fully participatory approach to develop research that has the ultimate goal of involving and engaging youth in the overall research process; from formulating the research goals, research questions, data collection, analysis of results, and translating data into practical solutions for the benefit of youth groups (Mirra et al., 2015).

For science communication projects that use YPAR and are interdisciplinary, or multidisciplinary, or even transdisciplinary (by combining social sciences, health, and environmental studies, among other disciplinary fields), theories of change can act as an epistemological foundation to guide the ethical procedures in research and practice. Moreover, for science communication, YPAR has been used as a methodological approach to engage high school students from multi-ethnic groups to be trained as co-researchers to develop community development initiatives, prevention campaigns, and dissemination activities. One such initiative was the *Summer Youth Research Institute* (SYRI), whose main goal was to create transformative projects that address the inequalities and issues that communities of color and Latinx immigrants faced in the northeastern United States. From 1997 to the early 2000s, the SYRI initiative was developed as a transformative approach to service-learning to support the education and active involvement of citizens in youth groups (Schensul & Berg, 2004).

In the age of social media and apps, YPAR has also been adapted to access new technologies, and different types of interaction but also combined with local knowledge to join together young people, by creating solutions for the social inequalities affecting them. In 2012, a multicultural group of researchers created the *Institute for Sustainable Economic Educational and Environmental Design* (ISEEED) in Oakland, California, as a technological tool to increase youth's the civic engagement to address the issue of food deserts and the massification of food resources of poor quality (i.e., fast food, liquor stores) among issues of urban gentrification that also leads to public health issues of obesity (Akom et al., 2016). ISEEED focuses on a community engagement strategy to develop collaborative research. Moreover, ISEEED uses a repository of information where people can participate by sharing information about practices for food access, food security, and alternatives of solutions to collaboratively build tools for social justice and health promotion.

To conduct YPAR, capacity-building is crucial to develop a study or an initiative collaboratively. One prior consideration to develop YPAR projects that follow ethical processes is to obtain bioethics and human-subjects review board authorizations. Moreover, signed consent and assent forms are necessary in the case of underage participants where they agree to participate as well as for their parents or legal guardians.

Arts-based Research

Participatory research and project tools have been also used to work in versatile ways with different audiences, not only to collect data and co-design but also to explore emancipatory practices, which support vulnerable people in expressing and overcoming trauma and elucidating desired-based solutions that improve their quality of life. Arts-based research (ABR) is a form of qualitative research that can be combined with other research paradigms (Barone & Eisner, 2011) such as quantitative, other qualitative tools, and participatory research methodologies.

ABR is a transdisciplinary approach that combines different art genres in research contexts. Specifically, ABR can be used in all or some phases of research for problem generation, content generation, analysis, interpretation, and representation of data (Chilton & Leavy, 2014; Leavy, 2017). From its epistemological foundations, ABR assumes that arts can be used to create and convey meaning around a specific topic (Leavy, 2017). Consequently, it can be used in projects to explore the lived experiences of a particular audience, or to create a new concept individually or in group collaborations.

ABR emerged in the 1990s as a formal research paradigm from creative arts therapy which has a therapeutic approach mostly coined in human psychology. Since then it has developed multiple subgenres such as visual methodologies, performance inquiry, image-based research, installation art-as-research, story-work research, and lyric inquiry (Clover, 2011).

ABR is also combined with participatory methodologies to approach unserved populations and vulnerable groups in research to examine the experiences and enhance the understanding of human conditions in delicate issues (i.e., gender violence, racial discrimination, and trauma healing projects). For underage participants, ABR can be also applied in projects for children to represent emotions, foster learning, and develop narratives that elucidate their lived experiences (Kirova & Emme, 2008; Mand, 2012). To support art-learning projects the ABR approach has been used as an interdisciplinary approach that unites art and scientific research to motivate creativity development and elucidating experiences.

Arts-based research presents several genres, among the main ones we find the following: literary genres (narrative inquiry, autoethnography, fiction-based research, and poetic inquiry), performative genres (a/r/tographic inquiry, ethnodrama, and play building), visual arts (arts-based visual research, drawing and painting, collage, installation art, and comics), and audiovisual arts (film, video-voice, and photovoice) (Leavy, 2017).

Among literary genres, we find narrative inquiry that, as a method draws upon the techniques of storytelling and writing to make meaning about people's lived experiences to communicate truthful stories about their human experiences (Leavy, 2020). Another literary arts-based research subgenre is autoethnography, also known as autobiography, which combines writing and research and explores multiple layers of personal experiences, as well as cultural background and it connects these layers to represent identity and self-perception (Ellis & Bochner, 2000). To explore emotions and experiences through an approach to better understand facts and fiction. Fiction-based research offers multiple possibilities to access through imagination the stories that can be inaccessible. Moreover, it invites people to self-reflect and represent their lived experiences by creating stories that represent real-life situations (Green, 2015) in projects that require creating more understanding of specific topics. However, this genre faces multiple critiques and concerns (1) the method struggles with scientific rigor and validity to represent truths due to its fictional components; (2) as a subjective form, the method cannot be assessed properly; and (3) reflectivity is difficult to achieve when working with different participants that use different representations (Nayebzadah, 2016). Consequently, we can recommend using fiction-based research as part of the data collection toolkit that combines different data collection tools, in order to be able to triangulate, saturate information, or complement findings in a more holistic approach. Finally, poetic inquiry as an ABR method invites people to represent and enhance discussions about a specific topic and build upon emergent topics through the craft of writing. Some authors have used poetic inquiry to combine social sciences research with social change efforts to represent marginalized and stigmatized identities (Faulkner, 2017).

Among the performative genres of arts-based research, a/r/tography is located in-between and in doing to define community, knowledge, and research that combines arts-related methodologies and scholarly writing through living inquiry (Irwin & De Cosson, 2004). A/r/tography positions the researcher as an artist and the artists as researchers by using different arts to represent life situations, elicit emotions, raise awareness about social issues, or seek alternative ways to portray reality (Gouzouasis, 2018). For example, the use of sculptures as a tool to advocate in the public space about gender violence (Roman et al., 2020), or songwriting in combination with ethnography to raise awareness about the difficulties music artists face in their labor (Bakan, 2014). A/r/tography needs to train participants in artistic skills such as painting, dancing, or performance in order to invite people to participate, but it also needs to be combined with rigorous theoretical backgrounds and other research tools to collect the data that is crafted through images, text or performance to be analyzed (Irwin, 2014).

Few studies have applied a/r/tography to science communication studies, although, art can be a very effective public engagement strategy to improve the understanding of nonexperts about complex data. Art has been used to communicate the importance of water ecology, and the efforts of scientists to develop research. In *The River Listening* project, scientists collaborated with visual artists to use microphotography and scientific imaging to engage the local community around scientific research about water resources by creating art exhibits and workshops for nonexperts to promote scientific learning about water preservation (Tyurina, 2022). Alternatives for science communication for public engagement can also become transdisciplinary projects by inviting artists to collaborate on outreach strategies.

As part of ABR's performative genres, ethnodrama is a research methodology and an art representation that creates scripted performances from interviews, field notes, journal entries, and media content to be presented as a specific topic's representation (Lawrence, 2015). Presenting results of research through a performance allows audiences to make meaning about a specific issue. Ethnodrama has been used as a research method and an advocacy tool to raise awareness about elementary school teachers' work conditions in low-income and marginalized communities (Dickens, 2011). As a research method, ethnodrama is written as a script that represents data results emergent from data collection tools; it's combined with a discussion that interprets core issues through scholarly research to evidence the causes, consequences, and possible solutions (Petersen, 2013).

Arts-based visual research is among the arts-based visual arts, and it develops through the use of images in research. As other ABR methods, arts-based visual research challenges the line between art and science (Holm et al., 2018). To develop ABR visual research can use several visual forms such as photographs, comics, diagrams, drawings, maps, and graffiti to be combined with research tools. Here visual arts can intervene to represent data, create visuals to analyze data and work collaboratively with audiences to develop for example paintings from the ethnographic data (Knowles & Cole, 2008).

As a sub-genre of arts-based visual research genres, comics emerge among literature and moving pictures (Berninger et al., 2010) and constitute a communication tool of popular culture through the use of images and text in storytelling (Talon, 2004). Comics can be used as a research method to analyze and read the stories behind graphics and text as tools that have semiotic content that evidences the phenomenon of perceptions and opinions regarding specific social issues or topics (Mustaqim & Zahar, n.d.). For example, comics have been used in transdisciplinary projects for urban, geography, and mobility studies to embrace a geo-critical approach to create more understanding of the human experiences in specific spaces through geographic novels to analyze the space and its influences on people (Peterle, 2021). Also, comics have been applied to critical inquiry in media studies to represent and amplify critical essays from communication researchers, by producing comics in printed and digital formats to create cross-media discussions among cartoonists and researchers in mass media (Chute & Jagoda, 2014).

In Latin America, the online and social media-based platform *Pictoline,* created by Eduardo Salles, a former advertising creative director (Ferrusquía, 2017), develops comics about complex scientific data, and topics of public interest. Through humor and quotidian experiences, people relate to these comics by using media reports and peer-reviewed research papers.

ABR also offers participatory audiovisual tools. Among these we find photovoice, which invites research participants to take photos with a specific theme, then they are invited to select the photos that speak the most to them to comment on them in group discussions (Sutton-Brown, 2014). Photovoice is a powerful tool to raise awareness in communities and policymakers, as it evidences the difficulties, problems, or lived experiences with pieces of evidence created from the research participants' perspective. Moreover, as a method, it collects multiple perspectives on the same issue from different people (Gubrium & Harper, 2016). Photovoice usually uses exhibits to present to the broader public the issues raised to create interest in the public opinion and policymakers to motivate social change and actions (Wang, 2006). For example, a PAR study developed by Alice McIntyre revealed the reality of working-class women living in Belfast to explore the relationship between place and identity of women in their everyday lives at the intersections of gender, class, and religion (2003). This study evidenced how Irish women in Belfast have been systematically excluded, economically disadvantaged, and marginalized by institutions in their everyday lives. This project provided opportunities for advocacy and creating a network for support among women (Mcintyre, 2003).

Video voice is also a PAR and arts-based research methodology that seeks to capture movement, audio, and sequential narratives to evidence people's lived experiences around a specific issue (Catalani et al., 2012). To conduct video voice studies in PAR studies, participants receive video cameras or participate through smartphones cameras to document a specific topic and narrate what is being observed. Usually, video voice pieces are compiled into films or documentaries that are contrasted and discussed with scientific sources (Gubrium & Harper, 2016). Video voice has been used to document natural disasters and their consequences, such as the post-Katrina in New Orleans (Catalani et al., 2012), and also the educational experiences of youth by creating music bands (Hamalainen, 2020).

Some ABR genres can be used as data collection tools for larger participatory projects. For example, they could be used to co-design a science communication initiative or evaluate a program and the experiences audiences have

Table 3.1. Arts-Based Research Genres

Literary Genres	Narrative inquiry	Draws upon storytelling and writing to make meaning about participants' lived experiences (Leavy, 2020).
	Autoethnography	Combines writing and research in order to invite participants to develop autobiographies that have multiple layers of personal experiences (Ellis & Bochner, 2000).
	Fiction-based research	Invites participants to create fiction stories that represent their lived experiences and elicit their emotions (Green, 2015). Due to the subjectivity of the method, it is recommended to combine it with other data collection tools.
	Poetic inquiry	Uses the craft of writing to represent emotions and lived experiences through poetry. It is a useful tool for social change inquiry and working with marginalized people (Faulkner, 2017).
Performative Genres	A/r/tographic inquiry	Uses collaborative work of researchers and artists to represent life situations and experiences to portray reality through art techniques or representations (Irwin, 2014).
	Ethnodrama	Uses scripts to develop a performance to represent data results that elucidate and make meaning about the causes and consequences of social issues (Lawrence, 2015).
Visual Arts	Arts-based visual research	Uses images such as photographs, comics, diagrams, drawings, maps, and graffiti to be combined with research tools to present data, elaborate analysis along with research methods of qualitative, quantitative, and participatory paradigms (Holm et al., 2018).
	Comics	Reads and analyzes the stories behind text and graphics. Moreover, comics have semiotic content regarding opinions and perceptions of social issues (Talon, 2004).
Audiovisual Arts	Videovoice	Compiles collaboratively the experiences and evidences about a specific issue by capturing movement and narratives (Gubrium & Harper, 2016).
	Photovoice	Raises awareness in communities and policymakers through collaborative documentation through picture taking of an event or experiences of people in order to raise awareness of communities and policymakers (Wang, 2006).

Developed by the authors.

around specific messages or activities (see table 3.1). Further on we present some concepts and procedures to use PAR tools and their different functions in science communication.

The following case study will present an example to apply arts-based methodologies into participatory initiatives that collect the lived experiences of people to raise awareness about the life of immigrants.

Case Study 4: Fotonovelas as ABR-PAR Research

In the last decades, Canada faced a wave of immigration from developing countries. Families from South America and refugees started their new lives in many cities in Canada. In this new context, their sons and daughters also began or followed their education in the Canadian school system. This study explored the lived experiences of immigrant pre-teenagers in adapting to a new life.

Anna Kirova and Michael Emme (2008) used *fotonovela* as a research tool to explore the experiences of immigrant pre-teenagers in Canada in an inner-city school. The authors organized a photo club to invite young participants to work in pairs to take pictures of specific places (e.g., related to their school life in cafeterias, playgrounds) related to their school life. Later, through prompt questions, they described the activities of each of their pictures accompanied them with quotes or phrases that captured what each meant for them and framed the photos into a narrative.

The researchers combined this approach with interviews, and group discussions to understand the stories, participants created and to frame the pictures into a story that shows their everyday life contexts such as playing a sport, and how these scenarios presented friendship, rejection, inclusion, and agency (Kirova & Emme, 2008).

Participatory Data Collection Tools

PAR projects require matching tools that foster interaction, interchange, and collaborative work. Moreover, these tools offer participants an inclusive approach to get involved in the data collection process and further analysis. PAR data collection tools offer versatile options to investigate and systematize data through visual representation, group analysis (Kramer et al., 2011), and art and technology-based platforms.

Participatory Diagramming

As a participatory research tool, participatory diagramming is used in group discussions, focus groups, workshops, and collaborative work sessions to facilitate dialogue and explore ideas, experiences, and opinions around a specific issue or subject (Kesby, 2000). PAR diagrams use familiar materials for the participants, such as paper posters, blackboards, Post-it notes, and markers. In rural areas, groups can, even use seeds, sticks, and drawings on the ground can be used as well in order to use materials that are available for the project or that are more familiar to the participants to make them comfortable enough to engage. PAR diagrams can use any format such as grids, Venn diagrams, or word clouds.

PAR Diagramming is a versatile tool that enables researchers to work with groups in an inclusive manner. As a tool, it can be easily adapted to the group's needs; for example, with illiterate participants, symbology can be used to represent specific options for answers. Also, written notes or cards can be useful to construct participant or co-researchers' responses or ideas (Kindon, 2007).

Amid the "new normal" context of COVID-19 and social distancing guidelines, with the recommendations not to gather in groups, technology can be of great help. For example, in our research in Ecuador during the pandemic, we used participatory workshops in several studies with children and adults by using the video conference platform *Zoom* to carry out sessions, and its screen sharing function allowed the presentation of PAR diagrams. Moreover, diagrams were collaboratively filled with participants through *Mentimeter.*[2] The PAR Diagrams and discussions were recorded for further review and analysis.

Layouts. PAR diagramming can use matrices, tree diagrams, flow diagrams, ranking scales, columns, word clouds, process arrows, and timelines, among others. Here, we must reflect on which topics will be explored, the questions that will be used, and the information we are looking to obtain in order to choose layouts that are suitable to conduct sessions and explore visually through participants' responses.

To create groups to work on PAR diagrams it is necessary to organize them homogeneously, as in focus groups. The ideal number of participants is five to seven people per facilitator to create a productive work dynamic. For example, characteristics such as gender, age, work, or any criteria that guide the projects (i.e., single fathers of school children) can serve to group people accordingly for discussions (Martí, 2017).

Case Study 5: Exploring a Sensitive Health Topic

Mike Kesby (2000) used PAR diagramming to develop a pilot study about HIV and gender relations in Zimbabwe. He used local materials (e.g., stones, sticks, rocks, and bottle caps of different colors) that represented different answers and a grid layout drawn on the floor to facilitate sessions. The session's facilitator used prompt questions to motivate interactions; people responded to the questions by placing their symbols on the answer options on the grid. Later, the facilitator and researchers in the group would "interview the diagram" by asking participants about their answers. All sessions were recorded in video and audio and the PAR diagrams created were photographed for further analysis.

As the main topic of inquiry implied a deep exploration of the sex lives and sexual issues of participants considered vulnerable groups, some topics required private and anonymous to be responses. So, for example, to respond to questions related to monogamy, use of condoms, or even gender violence in couples, PAR diagrams were placed in empty rooms and people would enter to vote and then gather around to talk about the responses without referring to themselves. The study evidenced which sex behaviors represented risks for women and co-created solutions or strategies to manage HIV-AIDS infection or prevent it.

Asset Mapping

Also known as Community Asset Mapping (Kerka, 2003), this participatory data collection tool documents tangible and intangible resources by allocating them places. As a research tool, it draws from "appreciative inquiry; the recognition of social capital in communities; participatory approaches to development which are based on principles of empowerment and ownership; collaborative economic development models that place priority on making the best use of a community's resource base . . ." (Mathie & Cunningham, 2003, p. 475).

Assets can be persons, physical structures (e.g., buildings or institutions), natural resources, businesses or informal organizations (e.g., grassroots groups, clubs) (Kerka, 2003), skills of individuals and communities, households, citizens, and institutions (Kretzmann & McKnight, 1993). As participatory-visual tools map also help researchers and practitioners to visualize and solve problems that other data collection tools probably cannot evidence clearly (Gubrium & Harper, 2016).

Asset mapping can be used in CBPR or PAR projects at diagnosis stages to assess the effectiveness and usefulness of prior projects or available resources before starting a study (Wallerstein et al., 2017) or practice-based projects. It is also a useful tool in the execution phases of collaborative research or projects, as asset maps also represent a form of community engagement through which communities have the opportunity to mobilize existing strengths and resources (Kramer et al., 2012).

How to conduct asset maps? Asset maps in CBPR or PAR studies do not need to use sophisticated tools necessarily.

Selecting the tools. We recommend considering the skills and knowledge of the co-researchers. So, for example, if a group has some knowledge of technological tools and internet access, basic training on the use of *Google*

Maps and Google Earth can be enough to use these platforms' collaborative options. However, if we are working with groups that have little or no access to the internet and technology, we can use collaborative printed maps, drawings, pictures of places, and written descriptions according to the type of study. For example, first, invite people to draw and list places guided by a goal and provide a brief explanation for their choices, and then discuss it with the group. A second step could be assigning symbology to the type of places (colors, shapes, icons) and then locating these on a collaborative printed map and making photographs for the registry.

Guiding the mapping sessions. Additionally, the development of clear prompt questions or topics of discussion for the sessions is key. For this purpose, semi-structured questions and setting a clear goal for the study and the meetings is necessary for the project team to avoid deviations.

In our research, we have used asset mapping in PAR studies, to allocate places, resources, and people that are considered key assets for the successful execution of health interventions and science communication initiatives for non-expert audiences. As a research tool, it can offer in-depth insights to understand people's perceptions about a place, as well as, how and why certain resources and places are related to positive or negative uses.

For example, in a study focused on science communication for public engagement, (see chapter 4 to review the overall case study) asset maps were used in a PAR-qualitative study designed by Vásquez-Guevara (2019) to co-design strategies with teenagers around the *Eat Smart to Play Hard*[3] health promotion program. In one of four PAR co-design workshops, young co-researchers explored places to develop interventions for children's physical activity in Albuquerque, New Mexico by using *Google Maps*.

A prompt question was used to guide and facilitate collaborative work. For example, one such question was: "*Which places are available for children and parents to do physical activity and play in the city?*" To answer this question, co-researchers drew developed drawings about places they thought as the best for children and families to take part in physical activity in Albuquerque. After making their drawings, the student collaborators were trained to develop a collaborative map in *Google Maps* locating the sites to analyze all of the answers. Frequent answers were community centers, parks, and dog parks.

Co-researchers mentioned that parents usually take children to fast food venues such as McDonald's and Burger King because these venues have playgrounds for customers (see figure 3.8). A surprising fact was that there were more than twice as many fast-food venues compared to the number of public spaces, such as parks and community centers. Therefore, those commercial venues were more accessible to parents than public spaces. Even more surprising, was that co-researchers acknowledged that public spaces could be great places to develop open events for children and families outdoors, such as free dancing lessons or family workout routines. However, there were doubts about how feasible organizing these events would be due to the increasing unsafety in public spaces, as parks in the city were frequently related to drug consumption and homelessness. Consequently, the interventions were designed to develop workouts at home or in schools, and public spaces were discarded.

Fanzines or Zines

This DIY, or do it yourself publication, has its origins in marginalized communities to record and elicit their stories, spread information, and organize advocacy initiatives of anarchist groups, feminist grassroots organizations, and LBGTQI+ groups (Jeppesen, 2012). Zines emerged in the United States in the 1960s as self-publication magazines (hence the abbreviation "Zines") of counter-cultural movements to disseminate ideas and values, by addressing topics of social interest such as music, human rights, and social movements (Duncombe, 2014). As an advocacy tool Zines were popularized by counterculture movements of underrepresented groups who were oppressed by dominant ideologies in order to evidence their stories of pain, marginalization, and the need for social change (French & Curd, 2022). Some scholars have used Zines for advocacy and activism, to raise awareness among others, such as decision-makers, and to advocate for a more inclusive policymaking (Guzzetti & Gamboa, 2004). They have also been used in scholarly work and

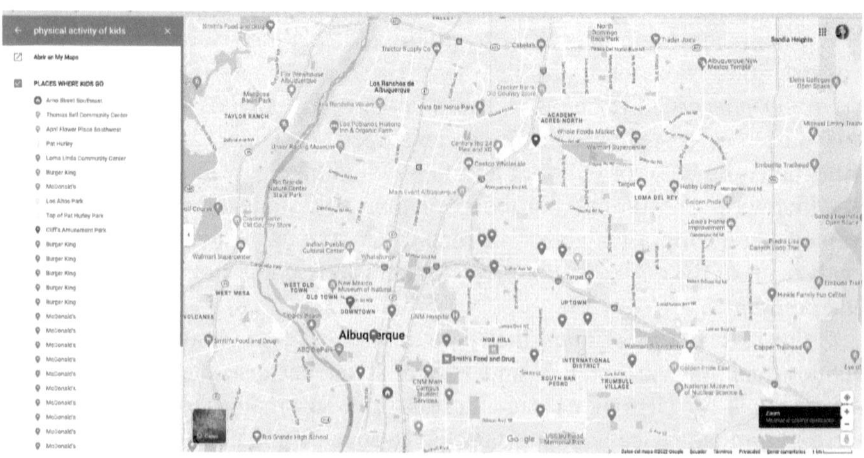

Figure 3.8. Asset map of spaces for children's physical activity and health information resource centers which was developed with the student collaborators. *Developed by Vásquez-Guevara (2019).*

in museums to make queer authors visible and engage the public in knowing their work (French & Curd, 2022).

However, currently Zines are used also in participatory projects to elicit lived experiences around specific topics by using simple tools such as paper, magazine cutouts, text, poetry, and drawings to present people's stories. As it uses people's skills and creativity it can be also considered an arts-based tool.

For science communication, we consider that Zines represent an opportunity to explore disseminating knowledge and co-design planned strategies for public engagement. In a recently published paper, Valli (2021) presented how participatory dissemination can bridge in-depth interviews and Zine-making to present data about representing the results of a gentrification study in the neighborhood of Bushwick in Brooklyn, New York City. The researcher worked with participants to create a handmade pamphlet, combining media texts and data to disseminate results to the broader community. The ultimate result was *The Bushwick Gentrification Zine* (Valli, 2021) which evidenced different kinds of conflict such as housing unaffordability, social disinvestment, and increased criminalization of communities of color.

Researchers of different fields could incorporate Zines in their public engagement strategies to elicit their ideas, suggestions for new inquiry, and science communication initiatives to engage broader audiences. As it is a versatile participatory tool, it can be adapted to any age group by engaging them in sessions that invite them to use their ideas and creativity to craft stories, drawings, text, and visual representations on paper about scientific content. Moreover, as it is a low-cost activity it also offers the opportunity to adapt to the available resources of research groups, schools, or community centers with minimum office or art supplies such as paper, scissors, color pens, recycled newspapers, or magazines.

Zine-making guidelines. To use Zines as a PAR tool in research or in co-design processes for science communication, we suggest the following guidelines.

1. Set a goal for the Zine to collect specific data, ideas, or suggestions. If we use Zines to develop ideas for events or communication strategies it is crucial to frame the main topic or goals. Offer templates or guidelines to the participants working individually—or in groups. Useful templates are questions or fill-in-the-blank statements. An example, of this could be: *best media to disseminate this message is . . . and its format should be . . .* (video, social media, news article).
2. Pick a format to develop the Zine. Frequently Zines use magazine sizes such as letter size A4 or A5 pages, or even smaller sizes, such as pocketbooks. It is necessary to adapt to the budget and available materials to construct the Zine; frequently Zines are handmade by cutting and

pasting paper sheets, making drawings, using collage techniques, and finally making copies to share them.

3. Select materials that are familiar to the audiences and easy to use for them, for example, paper sheets, colored pencils, and glue sticks, and which help audiences to present their ideas. For example, have enough recycled magazines or a computer with a printer to find images that are useful to present and illustrate how participants want to present their contributions.
4. Let creativity roll free; if participants have ideas to illustrate or present their ideas, or experiences through photographs, poetry, or any other printed or narrative form allow them enough freedom to create within the boundaries set for the project.

Citizen Science for Science Communication

As mentioned in chapter 1, citizen science and science communication are two intertwined disciplines that inform each other, but they also can work hand in hand around research projects or practical initiatives. In this section, we will analyze how to incorporate citizen science frameworks into research projects as well as present some cases of successful initiatives that have been useful to engage audiences and foster citizen collaboration with research groups.

As reviewed in chapter 1, citizen science projects emerged in the 1990s (Irwin, 2002), and these mostly happened through in-person meetings at science clubs, where workgroups were divided according to tasks guided by people's interests. However, novel citizen science projects, are not bound to in-person activities only; alternatively, these offer an open space for online participation through technological platforms or websites (Nov et al., 2014).

Planning Citizen Science Initiatives

Citizen science projects and science communication are disciplines that go hand in hand and enrich each other. Science communication is key to inviting people to become collaborators in citizen science projects, training researchers in public speaking skills, and communicating collaboration results between scientists and laypeople to the broader public. On the other hand, citizen science is a method itself that can be used in science communication projects to facilitate the co-creation of new knowledge and tools and resources that are useful for people to address specific issues. For example, for a co-creation or co-design project, participants or co-researchers could jointly develop resources for health COVID-19 prevention by developing tutorials about the correct mask use. Here researchers would source information about each type of mask, and participants or co-researchers could develop a narrative, and participate in the videos demonstrating the correct mask use and disposal.

At its planning stage, developing citizen science needs to question and reflect critically if the project will reach its goals by including participants or co-researchers, as well as consider how the project will benefit or not its participants of the project. At this point, it is necessary to determine if citizen science represents a correct and feasible approach for the project.

After defining if citizen science is a good approach to follow to develop a project, it is also necessary to define the scope and roles researchers and participants will have as co-researchers. Tweddle, Robinson, Pocock, and Roy (2012), in their guidebook *Guide to Citizen Science, Developing, Implementing, and Evaluating Citizen Science to Study Biodiversity and the Environment in the UK,* propose several questions to define key information to plan and define the project's functioning and processes.

- What geographic or temporal scale are you aiming to cover?
- How much data do you want to gather and analyze?
- Can volunteers help to gather and analyze these data?
- Are there other ways of gathering or analyzing the data?
- To whom will your project appeal?
- What might be their motivation for taking part?
- Can you support participants' involvement by providing training and coordination?
- Do you have the resources to develop and promote the project and share findings with participants?
- Are similar projects already in existence? It may be more efficient to add to existing schemes or work with other organizations than to set up a new project. (p. 7)

At this early stage, determining locations, timeframes, people's functions and researchers' functions, data collection, and reasons to motivate people's engagement will also be useful to establish. This will allow the planning team to allocate resources, develop a science communication strategy, and map prior projects that were developed that can provide useful information and guidelines to refine our project.

Successful citizen science projects work best when the following are present: (1) the project represents mutual benefit for all involved, researchers and participants; (2) it has clear goals that are communicated openly and publicly; (3) the researchers or experts have appropriate expertise in data collection, analysis, and science communication skills; (4) the project includes evaluation components and spaces to listen to ideas for improvement and adaptation; (5) trials or pretests are carried out to ensure effectiveness and the opportunity to correct errors; (6) they offer support to participants/co-researchers; (7) when skill-sets and motivations of all the parties involved are understood as diverse and tasks are assigned accordingly; (8) all team members feel included and understand the goals and relevance of their roles; (9) the project represents an enjoyable experience for all team members; (10) the data and knowledge that would be obtained is relevant and useful to all parties (Tweddle et al., 2012).

After defining if citizen science is a useful approach that is suitable for the type of project that will be conducted the processes and the steps through which these will operate and how these will function as well as, the tools that will be used need to be defined. For example, if co-researchers support the data collection process, we will need to define how the data will be reported (i.e., pictures uploaded to an online system).

Collaboration Incentives

As we mentioned before, in order to motivate participants′ engagement, projects need to offer incentives or compensations that are attractive and interesting enough to attract people's collaboration without conditioning their effective and ethical participation. Citizen science projects frequently attract people through learning opportunities, such as skill-based courses offered by educational institutions and, access to experts in fields of their personal interest.

However, available studies evidence that frequent motivations that attract people to citizen science projects as co-researchers include the opportunity to contribute to scientific research; learning more about topics that are useful for their education or professional career; the experiences of discovery where they appreciate new phenomena or the scientific explanations for topics of their interest; creating a community where co-researchers or participants can meet other people with similar interests; developing teaching tools and opportunities that can be useful resources for teaching and learning; enjoying the beauty of sciences through the study of people, nature, and events; fun experiences that are enjoyable to people; finally, enjoying to help projects that aim to improve living conditions and the advancement of science (Raddick et al., 2019).

So, for citizen science projects, we suggest incorporating activities that support networking among researchers, co-researchers and other people with similar interests. Some initiatives that can attract people are special events such as science cafés, webinars, and science clubs at places such as museums, restaurants, or even scientific laboratories where people gather frequently for Q & A sessions with experts or conduct an experiment with experts to learn more about topics of their interest.

For example, the California Academy of Sciences (n.d.) organizes family sleepovers at their headquarters in Golden Gate Park in San Francisco, California. This paid event invites children and their parents or guardians to sleep over at their museum and camp at the exhibit area and enjoy cookies and milk, dinner and breakfast, and explore the Academy's rainforest and planetarium. The attendees must take their own sleeping bag or camping tent.

Science cafés are another engagement format that either science communication initiatives or citizen science projects frequently implement. In the 1990s, this type of initiative first appeared in France as *cafe scientifique* to communicate science with the ultimate goal of improving audiences' interest and engagement (Clery, 2003). Science cafés are frequently considered informal learning activities for audiences of different ages and cultures done outside university classrooms or scientific events auditoriums (Honma, 2010). Instead, science cafés occur in informal settings that are attractive to audiences, such as cafés and restaurants that people can attend to learn more about topics of their interest.

Some of the frequent goals that science cafés have are informing citizens about a study´s scientific findings and the impact on their lives (Dijkstra, 2017). A recent study explored the audience's motivations to learn about science (Childers et al., 2021); among these the attendees revealed that their motivations were learning and gaining new knowledge that fulfills their personal needs and interests, as well as the opportunity to meet new people and have social interactions with others with their same interests to create a social network.

Organizing engaging activities can be difficult for scientists; however, we suggest some guidelines based on our prior studies (e.g., Vásquez-Guevara, 2019; White, 2013). To organize a science café, we suggest centering the event's focus on audiences, first, to investigate the habits and interests (i.e., places they like to attend, type of language they use, their main concerns and needs, and time and date preferences for events). For this purpose, a brief survey can be applied. We suggest doing in-depth interviews to explore more about their specific topics of interest and the implications that a specific issue might have for them. For example, if we are seeking to develop a science café about gender violence that affects females and audiences need resources they can use in these cases, then the event needs to present data about causes, consequences, and solutions and offer spaces for consultation with guest field experts. A second planning step would be to use the audience's information to select a place that is frequented by the target audience to organize the event and to select stakeholders that can bring resources and solutions to the audience. Third, we suggest developing scientific content that attends to real-life contexts such as daily life routines, or any daily life situation that is useful to frame the main topic and to problematize or argue about an issue. However, we do not want people to fall into panic, but to offer alternatives and solutions that are science-based and lead audiences to feasible actions. For example, if we are talking about healthy nutrition, we might focus on daily meals and fast food or unhealthy snacks, and how these can be replaced by healthier options. Finally, we suggest that scientists become the main speakers and organizers of science cafés or any science communication initiative. Studies show that people are usually interested in discipline experts in, and they will be more open to engaging and interacting with scientists than a trained public speaker (Besley et al., 2019; McCune, 2009).

From our experience, science cafés might be a great opportunity to build trust with communities and citizen organizations, but especially to listen to audiences and create more understanding about the complexity of the issues they face. Moreover, this is a great opportunity for researchers to get more ideas about future projects and research.

Project Execution

Once planned and ready to initiate the execution of a citizen science project, how people can participate in the citizen science initiative and the selection criteria for co-researchers needs to be defined. Again, at this point, an important point is to remember is that citizen science is a fully collaborative research framework where people are actively involved and responsible for the development of the research process. Knowledge is co-created between researchers and non-expert people taking part in a study. In citizen science, different from traditional Western scientific research frameworks where projects take a top-down approach from researchers to people, projects instead have a bottom-up framework where people address their local needs with the support of researchers through a network of collaboration that creates new knowledge to achieve real-world solutions (Hannibal, 2016).

As our citizen science project will develop through a collaboratory framework with non-experts, it is crucial to first develop a system and procedures that have accessible use and offer sufficient support materials, such as guides, instructions, frequently asked questions (FAQ), and mentoring spaces in workshops or meetings (Tweddle et al., 2012). Once these materials are ready, we suggest developing an initial training workshop where co-researchers and researchers meet to learn more about how to conduct the project, and the use of its tools, processes, and support materials. One key is to offer online support chats or messaging to do consultations or report any emergent situations.

As science communication professionals, we suggest first developing these materials by tailoring the contents and messages according to the project's needs and audiences' preferences; then pre-test them before initiating the project with a focus group of the co-researchers. This evaluation strategy will enable the team to refine and correct the materials before the overall execution.

Case Study 6 presents the National Geographic's Citizen Science initiative *Bioblitz and INaturalist* to explain how a citizen science initiative can function through online and mobile-based strategies to develop data collection registries of plants and animal species with people around the world.

Case Study 6: Bioblitz and INaturalist: a Discovery of the Natural World

National Geographic, INaturalist, and the California Academy of Sciences developed the mobile app *Seek* (National Geographic Society, n.d.). Children and adults of all ages can become citizen scientists by downloading the app, creating a personal account, and taking pictures of plants, bugs, and animals. Co-researchers can participate in monthly challenges that put them at different levels and will award them with digital badges that can be shared on social media. Underage co-researchers are encouraged to participate with their parents, family, or school teachers.

Seek operates through an informational website that provides clear instructions in different languages, a guidebook, and information about the monthly challenges people can join based on their interests or country.

This initiative feeds a worldwide registry at the Global Biodiversity Information Facility, an open-source database of species in short periods of time to evaluate collectively the biodiversity in urban and rural areas. Data is used as a resource used by scientists to undertake further studies and by policymakers around the world to develop environmental protection laws and programs (National Geographic Society, n.d.).

CONLCUSION

This chapter offered a broad exploration of how we can develop collaborative initiatives around science communication projects in different disciplines and topics. Moreover, through different participatory methods, we explored how these can be used as a paradigm to include science communication in grants and multidisciplinary projects through different tools and processes of evaluation and co-design.

The case studies presented here combine our very own projects and cases of other scholars and institutions that might offer you different ideas and guidelines to implement science communication research or practical initiatives in your own projects.

In chapter 4, we present two case studies and some specific guidelines to focus on designing science communication strategies. These cases explore how to conduct participatory projects. The first case presents how to design a science communication from a bottom-up approach to co-create science-based resources for audiences. The second case study presents a case to retool an existing program that wanted to explore how to build deeper engagement with its audiences to develop new strategies that invited them to actively participate in the program.

Finally, in chapter 4 we discuss the *Science for All* guidelines to offer a framework to develop effective strategies that engage audiences and foster meaningful collaborations between scientists and their audiences, science appreciation, and audience engagement.

NOTES

1. Translation developed by the authors from the book *Children of the Days* by Eduardo Galeano, from the passage *September 7, El Visitante*, September 7, The Visitor.

2. Mentimeter is a online-based platform and software, fully customizable, to build interactive presentations to engage audiences and that is fully customizable. For more information: www.mentimeter.com.

3. Eat Smart to Play Hard is a healthy-habits promotion school-based program developed by the UNM Health Sciences Prevention Research Center for New Mexico's elementary schools to prevent child obesity and chronic illnesses. For more information visit: https://hsc.unm.edu/medicine/departments/pediatrics/divisions/pps/initiatives/esph/.

BIBLIOGRAPHY

Akom, A., Shah, A., Nakai, A., & Cruz, T. (2016). Youth Participatory Action Research (YPAR) 2.0: How technological innovation and digital organizing sparked a food revolution in East Oakland. *International Journal of Qualitative Studies in Education: QSE*, *29*(10), 1287–1307. https://doi.org/10.1080/09518398.2016.1201609

Andrade, S., Lachat, C., Ochoa-Aviles, A., Verstraeten, R., Huybregts, L., Roberfroid, D., Andrade, D., Van Camp, J., Rojas, R., Donoso, S., Cardon, G., & Kolsteren, P. (2014). A school-based intervention improves physical fitness in Ecuadorian adolescents: a cluster-randomized controlled trial. *The International Journal of Behavioral Nutrition and Physical Activity*, *11*(1), 153. https://doi.org/10.1186/s12966-014-0153-5

Atkinson, R., & Mason, C. (2014). *Experiments in engagement: Review of literature around engagement with young people from disadvantaged backgrounds: April 2014*. Wellcome Trust.

Ayure, M., & Triana, R. (2022). Participatory science communication for transformation in Colombia. *Journal of Science Communication*, *21*(2), Y03. https://doi.org/10.22323/2.21020403

Bakan, D. L. (2014). *A song of songs: A/r/tography, autoethnography, and songwriting as music education research* [Doctoral Dissertation, University of British Columbia]. https://open.library.ubc.ca/collections/ubctheses/24/items/1.0167101

Ball, R., Duncanson, K., Ashton, L., Bailey, A., Burrows, T. L., Whiteford, G., Henström, M., Gerathy, R., Walton, A., Wehlow, J., & Collins, C. E. (2021). Engaging new parents in the development of a peer nutrition education model using participatory action research. *International Journal of Environmental Research and Public Health*, *19*(1), 102. https://doi.org/10.3390/ijerph19010102

Bammer, G., O'Rourke, M., O'Connell, D., Neuhauser, L., Midgley, G., Klein, J. T., Grigg, N. J., Gadlin, H., Elsum, I. R., Bursztyn, M., Fulton, E. A., Pohl, C., Smithson, M., Vilsmaier, U., Bergmann, M., Jaeger, J., Merkx, F., Vienni Baptista, B., Burgman, M. A., . . . Richardson, G. P. (2020). Expertise in research integration and implementation for tackling complex problems: when is it needed, where can it be found and how can it be strengthened? *Palgrave Communications*, *6*(1), 1–16. https://doi.org/10.1057/s41599-019-0380-0

Barone, T., & Eisner, E. W. (2011). *Arts Based Research*. SAGE Publications. https://play.google.com/store/books/details?id=DJfcs6igLFAC

Bartholomew, L. K., Parcel, G. S., Kok, G., Gottlieb, N. H., & Fernandez, M. E. (2011). *Planning health promotion programs: an intervention mapping approach*. John Wiley & Sons.

Berninger, M., Ecke, J., & Haberkorn, G. (2010). *Comics as a nexus of cultures: Essays on the interplay of media, disciplines and international perspectives (Critical explorations in science fiction and fantasy)*. McFarland & Company Inc. Publishers.

Besley, J. C., O'Hara, K., & Dudo, A. (2019). Strategic science communication as planned behavior: Understanding scientists' willingness to choose specific tactics. *PloS One*, *14*(10), e0224039. https://doi.org/10.1371/journal.pone.0224039

Bevan, B. B., Barton, A. C., & Garibay, C. (2018). Why broaden perspectives on broadening participation in STEM. *Center for Advancement of Informal Science Education*. https://www.informalscience.org/sites/default/files/BP-1-Why-Broaden-Perspectives.pdf

Bonney, R. (2021). Expanding the impact of citizen science. *Bioscience*, *71*(5), 448–451. https://doi.org/10.1093/biosci/biab041

Bowater, L., & Yeoman, K. (2013). *Science communication*. Wiley-Blackwell.

Brossard, C., Feuillet, V., Schmitt, A., Randriamampita, C., Romao, M., Raposo, G., & Trautmann, A. (2005). Multifocal structure of the T cell-dendritic cell synapse. *European Journal of Immunology*, *35*(6), 1741–1753. https://doi.org/10.1002/eji.200425857

Brossard, D., & Scheufele, D. A. (2013). Science, new media, and the public. *Science*, *339*(6115), 40–41. https://doi.org/10.1126/science.1232329

Bucchi, M., & Trench, B. (Eds.). (2014). *Handbook of public communication of science and technology*. Routledge.

Bucchi, M., & Trench, B. (Eds.). (2021). *Routledge handbook of public communication of science and technology: Third edition*. Routledge. https://doi.org/10.4324/9781003039242

California Academy of Sciences. (n.d.). *Penguins+pajamas sleepovers*. California Academy of Sciences. Retrieved May 21, 2022, from https://www.calacademy.org/penguinspajamas-sleepovers

Cammarota, J. (2017). Youth participatory action research: A pedagogy of transformational resistance for critical youth studies. *Journal for Critical Education Policy Studies (JCEPS)*, *15*(2). http://www.jceps.com/wp-content/uploads/2017/10/15-2-7.pdf

Campos, R. (2022). Including younger children in science-related issues using participatory and collaborative strategies: A pilot project on urban biodiversity. *Journal of Science Communication*, *21*(2), N07. https://doi.org/10.22323/2.21020807

Catalani, C. E. C. V., Veneziale, A., Campbell, L., Herbst, S., Butler, B., Springgate, B., & Minkler, M. (2012). Videovoice: community assessment in post-Katrina New Orleans. *Health Promotion Practice*, *13*(1), 18–28. https://doi.org/10.1177/1524839910369070

Cavestro, L. (2003). PRA-participatory rural appraisal concepts methodologies and techniques. *Padova University. Padova PD. Italia*. https://www.researchgate.net/profile/Saswatik-Tripathy/post/Can_someone_suggest_a_preferred_method_of_measuring_coastal_livelihood_sustainability/attachment/5e14cb55cfe4a777d401e878/AS%3A844773380349952%401578421077767/download/participatory+rural+appraisal.pdf

Chambers, R. (1994). The origins and practice of participatory rural appraisal. *World Development*, *22*(7), 953–969. https://doi.org/10.1016/0305–750X(94)90141–4

Chandra, G. (2010). Participatory rural appraisal. In Katiha, P. K., Vaas, K. K., Sharma, A. P., Bhaumik, U., & Chandra, G. (Eds.), *Issues and tools for social science research in inland fisheries*. Central Inland Fisheries Research Institute, Barrackpore, Kolkata, India. Bulletin, 163, 286–302. https://www.researchgate.net/profile/Ganesh-Chandra-5/publication/265127750_Participatory_Rural_Appraisal/links/540474e30cf23d9765a63588/Participatory-Rural-Appraisal.pdf

Chevalier, J., & Buckles, D. (2013). *Participatory action research: Theory and methods for engaged inquiry*. Routledge. https://doi.org/10.4324/9780203107386

Chevalier, J., & Buckles, D. (2019). *Participatory action research: Theory and methods for engaged inquiry* (2nd ed.). Routledge. https://www.taylorfrancis.com/books/mono/10.4324/9781351033268/participatory-action-research-jacques-chevalier-daniel-buckles

Childers, G., Governor, D., Osmond, D., & Britton, S. (2021). Science cafés: Exploring adults' motivation to learn science in a community space. *Research in Science Education, 52,* 1055–1073. https://doi.org/10.1007/s11165-020-09982-2

Chilton, G., & Leavy, P. (2014). Arts-based research practice: Merging social research and the creative arts. In P. Leavy (Ed.), *The Oxford Handbook of Qualitative Research* (pp. 403–422). https://books.google.com/books?hl=en&lr=&id=RTaTAwAAQBAJ&oi=fnd&pg=PA403&dq=Leavy+arts+based+research&ots=AK376Aa1aE&sig=LYM3UlF_5aHtuFT79d9VErCXqKQ

Chiu, L. F. (2008). Health promotion and participatory action research: The significance of participatory praxis in developing participatory health intervention—SAGE Research Methods. In P. Reason & H. Bradbury (Eds.), *The Sage handbook of action research*. https://methods.sagepub.com/base/download/BookChapter/the-sage-handbook-of-action-research/d47.xml

Chute, H., & Jagoda, P. (2014). Special issue: Comics & media. *Critical Inquiry, 40*(3), 1–10. https://doi.org/10.1086/677316

Clery, D. (2003). Earth observation. Satellite start-up offers cut-price window on the world. *Science*, *301*(5641), 1829–1831. https://doi.org/10.1126/science.301.5641.1829a

Clover, D. (2011). Successes and challenges of feminist arts-based participatory methodologies with homeless/street-involved women in Victoria. *Action Research*, *9*(1), 12–26. https://doi.org/10.1177/1476750310396950

Collins, K., Shiffman, D., & Rock, J. (2016). How are scientists using social media in the workplace? *PloS One*, *11*(10), e0162680. https://doi.org/10.1371/journal.pone.0162680

Constant, N., & Roberts, L. (2017). Narratives as a mode of research evaluation in citizen science: understanding broader science communication impacts. *Journal of Science Communication*, *16*(4), A03. https://doi.org/10.22323/2.16040203

Cooper, C. (2016). *Citizen science*. Overlook Press.

Crettaz von Roten, F. (2011). Gender differences in scientists' public outreach and engagement activities. *Science Communication*, *33*(1), 52–75. https://doi.org/10.1177/1075547010378658

Davies, S. R. (2008). Constructing communication. In *Science Communication, 29*(4), 413–434. https://doi.org/10.1177/1075547008316222

Davies, S. R. (2019). Science communication as emotion work: Negotiating curiosity and wonder at a science festival. *Science as Culture*, *28*(4), 538–561.

Dawson, E. (2018). Reimagining publics and (non) participation: Exploring exclusion from science communication through the experiences of low-income, minority ethnic groups. *Public Understanding of Science*, *27*(7), 772–786. https://doi.org/10.1177/0963662517750072

De Vos, P., Guerra, M., Sosa, I., del R Ferrer, L., Rodríguez, A., Bonet, M., Lefevre, P., & Van der Stuyft, P. (2011). Comprehensive participatory planning and evaluation (CPPE). *Social Medicine (Social Medicine Publication Group)*, *6*(2), 106–117. http://www.socialmedicine.info/index.php/socialmedicine/article/view/593

Dickens, G. R. (2011). Down the rabbit hole: Toward appropriate discussion of methane release from gas hydrate systems during the Paleocene-Eocene thermal maximum and other past hyperthermal events. *Climate of the Past*, *7*(3), 831–846. https://cp.copernicus.org/articles/7/831/2011/

Dijkstra, A. (2017). Analysing Dutch Science Cafés to better understand the science-society relationship. *Journal of Science Communication*, *16*(1), A03. https://doi.org/10.22323/2.16010203

Duncombe, S. (2014). *Notes from underground: Zines and the politics of alternative culture*. Microcosm Publishing. https://play.google.com/store/books/details?id=n-93CAAAQBAJ

Dunwoody, S. (2014). Science journalism: Prospects in the digital age. In M. Bucchi & B. Trench (Eds.), *Handbook of public communication of science and technology* (pp. 43–55). Routledge.

Eelderink, M., Vervoort, J., & van Laerhoven, F. (2020). Using participatory action research to operationalize critical systems thinking in social-ecological systems. *Ecology and Society*, *25*(1). https://doi.org/10.5751/ES-11369–250116

Ellis, C., & Bochner, A. (2000). Autoethnography, personal narrative, reflexivity: Researcher as subject. In N. K. Denzin & Y. S. Lincoln (Eds.), *Handbook of qualitative research* (2nd Ed.) (pp. 733–768). Sage Publications. https://scholarcommons.usf.edu/spe_facpub/91/

Enros, K., & Bandelli, A. (2018). Beyond self-confidence: A participatory evaluation of personal change in Science Gallery's mediators. *Journal of Science Communication*, *17*(3), N01. https://doi.org/10.22323/2.17030801

Faulkner, S. L. (2017). Poetic inquiry: Poetry as/in/for social research. In P. Leavy (Ed.), *The handbook of arts-based research* (pp. 208–230). Guilford Press.

Fernández-Aballí Altamirano, A. (2020). The importance of Paulo Freire to communication for development and social change. In J. Servaes (Ed.), *Handbook of communication for development and social change* (pp. 309–327). Springer Singapore. https://doi.org/10.1007/978-981-15-2014-3_76

Ferrusquía, R. M. (2017, February 22). Pictoline: diseño de información para un mundo disperso. *Forbes Magazine*. https://www.forbes.com.mx/pictoline-diseno-informacion-mundo-disperso/

Fine, M. (2009). Postcards from metro America: Reflections on youth participatory action research for urban justice. *The Urban Review*, *41*(1), 1–6. https://doi.org/10.1007/s11256-008-0099-5

Fischhoff, B. (2013). The sciences of science communication. *Proceedings of the National Academy of Sciences, 110*(3_suppl.), 14033–14039. https://doi.org/10.1073/pnas.1213273110

Fox, B. A. (1987). Interactional reconstruction in real-time language processing. *Cognitive Science*, *11*(3), 365–387. https://doi.org/10.1016/S0364-0213(87)80011–3

Freire, P. (2000). *Pedagogy of the oppressed* (M. Bergman Ramos, Trans.; 30th ed.). Continuum International Publishing Group.

French, J., & Curd, E. (2022). Zining as artful method: Facilitating zines as participatory action research within art museums. *Action Research*, *20*(1), 77–95. https://doi.org/10.1177/14767503211037104

Galeano, E. (2013). *Children of the days: A calendar of human history*. Bold Type Books.

Gouzouasis, P. (2018). A/r/tographic inquiry in a new totality. In P. Leavy (Ed.), *The handbook of arts-based research* (pp. 233–246). *Guilford Press.*

Green, D. (2015). Exploring and sharing arts therapy experiences through fiction-based research. *Australian and New Zealand Journal of Art Therapy (ANZJAT)*, *10*(1), 33–42. https://www.jocat-online.org/s/7-ANZJAT-2015-DG.pdf

Gubrium, A., & Harper, K. (2016). *Participatory visual and digital methods*. Routledge. https://doi.org/10.4324/9781315423012

Guzzetti, B. J., & Gamboa, M. (2004). Zines for social justice: Adolescent girls writing on their own. *Reading Research Quarterly*, *39*(4), 408–436. https://doi.org/10.1598/rrq.39.4.4

Habermas, J. (1985). *The theory of communicative action: Volume 2: Lifeword and system: A critique of functionalist reason.* Beacon Press.

Habermas, J. (1991). *The structural transformation of the public sphere: An inquiry into a category of bourgeois society*. MIT Press.

Hacker, K. (2013). *Community-based participatory research*. SAGE Publications. https://play.google.com/store/books/details?id=VTh1AwAAQBAJ

Hacker, K., Tendulkar, S. A., Rideout, C., Bhuiya, N., Trinh-Shevrin, C., Savage, C. P., Grullon, M., Strelnick, H., Leung, C., & DiGirolamo, A. (2012). Community capacity building and sustainability: outcomes of community-based participatory research. *Progress in Community Health Partnerships: Research, Education, and Action*, *6*(3), 349–360. https://doi.org/10.1353/cpr.2012.0048

Hamalainen, J. (2020). *Videovoice perspectives of beginning band students at home* [Master's Thesis, Dominican University of California]. https://doi.org/10.33015/dominican.edu/2020.EDU.09

Hand, C., Rudman, D. L., McGrath, C., Donnelly, C., & Sands, M. (2019). Initiating participatory action research with older adults: Lessons learned through reflexivity. *Canadian Journal on Aging = La Revue Canadienne Du Vieillissement*, *38*(4), 512–520. https://doi.org/10.1017/S0714980819000072

Hannibal, M. E. (2016). *Citizen scientist: Searching for heroes and hope in an age of extinction*. The Experiment. https://play.google.com/store/books/details?id=2mB3CwAAQBAJ

Haywood, B. K., & Besley, J. C. (2014). Education, outreach, and inclusive engagement: Towards integrated indicators of successful program outcomes in participatory science. *Public Understanding of Science*, *23*(1), 92–106. https://doi.org/10.1177/0963662513494560

Hecker, S., Luckas, M., Brandt, M., Kikillus, H., Marenbach, I., Schiele, B., Sieber, A., van Vliet, A. J. H., Walz, U., & Wende, W. (2018). Stories can change the world—citizen science communication in practice. In S. Hecker, M. Haklay, A. Bowser, Z. Makuch, J. Vogel, & A. Bonn (Eds.), *Citizen science: Innovation in open science, society and policy* (pp. 445–462). https://doi.org/10.2307/j.ctv550cf2.37

Herr, K., & Anderson, G. L. (2014). *The action research dissertation: A guide for students and faculty*. SAGE Publications. https://play.google.com/store/books/details?id=OsnBAwAAQBAJ

Holbrook, J. B. (2010). The use of societal impacts considerations in grant proposal peer review: A comparison of five models. *Technology & Innovation*, *12*(3), 213–224.

Holliman, R., Whitelegg, E., Scanlon, E., Smidt, S., & Thomas, J. (2008). Models of science communication. In R. Holliman, E. Whitelegg, E. Scanlon, S. Smidt, & J. Thomas (Eds.), *Investigating science communication in the Information Age: Implications for public engagement and popular media* (p. 288). Oxford University Press; Open University Press. http://shura.shu.ac.uk/id/eprint/2816

Holm, G., Sahlström, F., & Zilliacus, H. (2018). Arts-based visual research. In P. Leavy (Ed.), *The handbook of arts-based research* (pp. 311–335). Guilford Press.

Honma, Y. (2010). SCCJ cafe (2): Science cafe and scientific communication. *Journal of Computer Chemistry Japan*, *9*(4), A13–A16. https://doi.org/10.2477/jccj.h2215

Horst, M., & Michael, M. (2011). On the shoulders of idiots: Re-thinking science communication as "event." *Science as Culture*, *20*(3), 283–306. https://doi.org/10.1080/09505431.2010.524199

Humm, C., & Schrögel, P. (2020). Science for all? Practical recommendations on reaching underserved audiences. *Frontiers in Communication*, *5*. https://doi.org/10.3389/fcomm.2020.00042

Irwin, H. (2002). Is scientific investigation of postmortem survival an anachronism?: The demise of the survival hypothesis. *Australian Journal of Parapsychology*, *2*(1), 19–27.

Irwin, R. (2014). Turning to a/r/tography. *Journal of Research in Art Education*, *15*(1), 21–40. https://doi.org/10.20977/kkosea.2014.15.1.21

Irwin, R., & De Cosson, A. (2004). *A/r/tography: Rendering self through arts-based living inquiry*. Pacific Educational Press.

Israel, B. A., Lantz, P. M., McGranaghan, R. J., Guzman, J. R., Lichtenstein, R., & Rowe, Z. (2005). Documentation and evaluation of CBPR partnerships. In B. A. Israel, E. Eng, A. J. Schulz, & E. A. Parker (Eds.), *Methods in community-based participatory research* (pp. 255–277). https://books.google.com/books?hl=en&lr=&id=fN6I3_IJW-wC&oi=fnd&pg=PA255&dq=Israel+2003+CBPR&ots=HZuvh6Dxfl&sig=dm6vLpXbv8l4xma7krlESIgOsDM

Jensen, E. (2014). The problems with science communication evaluation. *Journal of Science Communication*, *13*(1), C04. https://doi.org/10.22323/2.13010304

Jeppesen, S. (2012). DIY zines and direct-action activism. In K. Kozolanka, P. Mazepa, & D. Skinner (Eds.), *Alternative media in Canada* (pp. 264–281). UBC Press.

Kalmár, É., & Stenfert, H. (2020). Science Communication as a design challenge in transdisciplinary collaborations. *Journal of Science Communication, 19*(4), C01. https://jcom.sissa.it/archive/19/04/JCOM_1904_2020_C01

Kerka. (2003). Community asset mapping. *Trends and Issues Alert*, *47*(1). https://eric.ed.gov/?id=ED481324

Kesby, M. (2000). Participatory diagramming: deploying qualitative methods through an action research epistemology. *Area*, *32*(4), 423–435. https://doi.org/10.1111/j.1475–4762.2000.tb00158.x

Kindon, S., Pain, R., & Kesby, M. (Eds). (2007). *Participatory action research approaches and methods: Connecting people, participation and place.* Routledge. https://doi.org/10.4324/9780203933671

Kirova, A., & Emme, M. (2008). Fotonovela as a research tool in image-based participatory research with immigrant children. *International Journal of Qualitative Methods*, *7*(2), 35–57. https://doi.org/10.1177/160940690800700203

Knowles, J., & Cole, A. (2008). *Handbook of the arts in qualitative research: Perspectives, methodologies, examples, and issues*. Sage. https://doi.org/10.4135/9781452226545

Kramer, J., Kramer, J., García-Iriarte, E., & Hammel, J. (2011). Following through to the end: The use of inclusive strategies to analyse and interpret data in participatory action research with individuals with intellectual disabilities. *Journal of Applied Research in Intellectual Disabilities: JARID*, *24*(3), 263–273. https://doi.org/10.1111/j.1468-3148.2010.00602.x

Kramer, S., Amos, T., Lazarus, S., & Seedat, M. (2012). The philosophical assumptions, utility and challenges of asset mapping approaches to community engagement. *Journal of Psychology in Africa*, *22*(4), 537–544. https://doi.org/10.1080/14330237.2012.10820565

Kretzmann, J., & McKnight, J. (1993). *Building communities from the inside out: A path toward finding and mobilizing a community's assets*. ACTA Publications.

Kurath, M., & Gisler, P. (2009). Informing, involving or engaging? Science communication, in the ages of atom-, bio- and nanotechnology. In *Public Understanding of Science, 18*(5), 559–573. https://doi.org/10.1177/0963662509104723

Lake, D., & Wendland, J. (2018). Practical, epistemological, and ethical challenges of participatory action research: A cross-disciplinary review of the literature. *Journal of Higher Education Outreach and Engagement*, *22*(3), 11–42. https://ojs01.galib.uga.edu/jheoe/article/view/1399

Lawrence, R. L. (2015). Dancing with the data: Arts-based qualitative research. In V. Wang (Ed.), *Handbook of research on scholarly publishing and research methods* (pp. 141–154). IGI Global. https://doi.org/10.4018/978-1-4666-7409-7.ch008

Leach, J., Yates, S., & Scanlon, E. (2008). Models of science communication. In R. Holliman, E. Whitelegg, E. Scanlon, S. Smidt, & J. Thomas (Eds.) *Investigating science communication in the Information Age: Implications for public engagement and popular media* (pp. 128–146). Oxford University Press; Open University Press.

Leavy, P. (2017). *Handbook of arts-based research*. Guilford Publications. https://play.google.com/store/books/details?id=fRYuDwAAQBAJ

Leavy, P. (2020). *Method meets art, third edition: Arts-based research practice*. Guilford Publications. https://play.google.com/store/books/details?id=sb_nDwAAQBAJ

Lee, N. M., VanDyke, M. S., & Cummins, R. G. (2018). A missed opportunity? NOAA's use of social media to communicate climate science. *Environmental Communication*, *12*(2), 274–283. https://doi.org/10.1080/17524032.2016.1269825

Leitch, A. (2022). Participatory science communication needs to consider power, place, pain and "poisson": A practitioner insight. *Journal of Science Communication*, *21*(02), N01. https://doi.org/10.22323/2.21020801

Lepori, B., & Greco, S. (2019). 18 Grant proposal writing as a dialogic process. *Science Communication*, *17*, 377.

Lewin, K. (1946). Action research and minority problems. *Journal of Social Issues*, *2*(4), 34–46.

Lloyd, R., Neilson, R., King, S., & Dyball, M. (2012). *Review of informal science learning*. Wellcome Trust.

Mand, K. (2012). Giving children a "voice": Arts-based participatory research activities and representation. *International Journal of Social Research Methodology, 15*(2), 149–160. https://doi.org/10.1080/13645579.2012.649409

Marschalek, I., & Schrammel, M. (2017). *Soziale Inklusion durch und in Wissenschaftskommunikation. Studie zur Situation von marginalisierten Kindern und Jugendlichen in Wien*. Centre for Social Innovation.

Martí, J. (2017). *La investigación—acción participativa: estructura y fases* [Participatory action research. Structure and phases]. [Doctoral Dissertation, Universidad Complutense de Madrid].

Massarani, L., & Merzagora, M. (2014). Socially inclusive science communication. *JCOM: Journal of Science Communication, 13*. https://go.gale.com/ps/i.do?id=GALE%7CA378248995&sid=googleScholar&v=2.1&it=r&linkaccess=abs&issn=18242049&p=AONE&sw=w

Mathie, A., & Cunningham, G. (2003). From clients to citizens: Asset-based community development as a strategy for community-driven development. *Development in Practice, 13*(5), 474–486. https://doi.org/10.1080/0961452032000125857

McCune, V. (2009). Final year biosciences students' willingness to engage: teaching–learning environments, authentic learning experiences and identities. *Studies in Higher Education, 34*(3), 347–361. https://doi.org/10.1080/03075070802597127

Mcintyre, A. (2003). Through the eyes of women: Photovoice and participatory research as tools for reimagining place. *Gender, Place and Culture: A Journal of Feminist Geography 10*(1), 47–66. https://doi.org/10.1080/0966369032000052658

McNiff, J. (2016). *You and your action research project*. Routledge. https://www.taylorfrancis.com/books/mono/10.4324/9781315693620/action-research-project-jean-mcniff

McNiff, J., & Whitehead, J. (2009). *Doing and writing action research*. SAGE. https://play.google.com/store/books/details?id=ApFzPt7uBCoC

Minkler, M. (2004). Ethical challenges for the "outside" researcher in community-based participatory research. *Health Education & Behavior: The Official Publication of the Society for Public Health Education, 31*(6), 684–697. https://doi.org/10.1177/1090198104269566

Mirra, N., Garcia, A., & Morrell, E. (2015). *Doing youth participatory action research: Transforming inquiry with researchers, educators, and students*. Routledge. https://doi.org/10.4324/9781315748047

Musoba, G., Jacob, S., & Robinson, L. (2015). The institutional review board (IRB) and faculty: Does the IRB challenge faculty professionalism in the social sciences? *The Qualitative Report*. https://doi.org/10.46743/2160–3715/2014.1413

Mustaqim, K., & Zahar, I. (n.d.). *Before methodology: In case of visual art-based research in studying comics art*. https://digilib.esaunggul.ac.id/public/UEU-Article-23116-5_0816.pdf

National Geographic Society. (n.d.). *BioBlitz program*. Retrieved May 21, 2022, from https://www.nationalgeographic.org/projects/bioblitz/

Nayebzadah, R. (2016). The truth behind fiction-based research. *Journal of Humanistic and Social Studies*, 7(2), 49–61. https://www.ceeol.com/search/article-detail?id=618262

Nisbet, M. C., & Scheufele, D. A. (2009). What's next for science communication? Promising directions and lingering distractions. *American Journal of Botany*, *96*(10), 1767–1778. https://doi.org/10.3732/ajb.0900041

Nov, O., Arazy, O., & Anderson, D. (2014). Scientists@Home: What drives the quantity and quality of online citizen science participation? *PloS One*, *9*(4), e90375. https://doi.org/10.1371/journal.pone.0090375

Ochoa-Avilés, A., Andrade, S., Huynh, T., Verstraeten, R., Lachat, C., Rojas, R., Donoso, S., Manuel-y-Keenoy, B., & Kolsteren, P. (2012). Prevalence and socioeconomic differences of risk factors of cardiovascular disease in Ecuadorian adolescents. *Pediatric Obesity*, *7*(4), 274–283. https://doi.org/10.1111/j.2047–6310.2012.00061.x

Ochoa-Avilés, A., Verstraeten, R., Huybregts, L., Andrade, S., Van Camp, J., Donoso, S., Ramírez, P. L., Lachat, C., Maes, L., & Kolsteren, P. (2017). A school-based intervention improved dietary intake outcomes and reduced waist circumference in adolescents: A cluster randomized controlled trial. *Nutrition Journal*, *16*(1), 79. https://doi.org/10.1186/s12937-017-0299-5

Ochoa-Avilés, A., Verstraeten, R., Lachat, C., Andrade, S., Van Camp, J., Donoso, S., & Kolsteren, P. (2014). Dietary intake practices associated with cardiovascular risk in urban and rural Ecuadorian adolescents: A cross-sectional study. *BMC Public Health*, *14*(1), 939. https://doi.org/10.1186/1471-2458-14-939

Omondi, L. A. (2020). Learning together: Participatory rural appraisal for coproduction of climate change knowledge. *Action Research*, 1476750320905901. https://doi.org/10.1177/1476750320905901

Orthia, L. A., McKinnon, M., Viana, J. N., & Walker, G. (2021). Reorienting science communication towards communities. *Journal of Science Communication*, *20*(3), A12. https://doi.org/10.22323/2.20030212

Ottawa Charter for Health Promotion (1986). World Health Organization—Europe. https://www.euro.who.int/__data/assets/pdf_file/0004/129532/Ottawa_Charter.pdf

Pearce, W., Brown, B., Nerlich, B., & Koteyko, N. (2015). Communicating climate change: Conduits, content, and consensus. *Wiley Interdisciplinary Reviews. Climate Change*, *6*(6), 613–626. https://doi.org/10.1002/wcc.366

Peterle, G. (2021). *Comics as a research practice: Drawing narrative geographies beyond the frame*. Routledge. https://www.taylorfrancis.com/books/mono/10.4324/9781003058069/comics-research-practice-giada-peterle

Petersen, E. B. (2013). Cutting edge(s). *Cultural Studies ↔ Critical Methodologies*, *13*(4), 293–298. https://doi.org/10.1177/1532708613487876

Pictoline. (2022, May 2). *Viernes · 18 · Febrero · 2022*. https://www.pictoline.com/timeline/2022/02/18/13hrs26min09sec?search=covid

Porter, L. (2010). *Unlearning the colonial cultures of planning*. Ashgate Publishing, Ltd. https://play.google.com/store/books/details?id=rGLKlRHpGcEC

Raddick, M. J., Prather, E. E., & Wallace, C. S. (2019). Galaxy zoo: Science content knowledge of citizen scientists. *Public Understanding of Science*, *28*(6), 636–651. https://doi.org/10.1177/0963662519840222

Riedlinger, M., Metcalfe, J., Baram-Tsabari, A., Entradas, M., Joubert, M., & Massarani, L. (2019). Telling stories in science communication: Case studies of scholar-practitioner collaboration. *Journal of Science Communication, 18*(5), N01. https://eprints.qut.edu.au/203716

Rocha, R., & Hua, F. (2018). Tackle all forms of under-representation in science. *Nature*, *557*(7707), 637–637. https://doi.org/10.1038/d41586-018-05289-5

Roman, L. G., del Pieve Gobbi, R., Beaudry, J.-S., & Blackbridge, P. (2020). Rippling excesses: A/r/tography becoming dis/a/r/tography. *International Journal of Education Through Art*, *16*(1), 63–82. https://doi.org/10.1386/eta_00017_1

Schensul, J. J., & Berg, M. (2004). Youth Participatory Action Research: A transformative approach to service-learning. *Michigan Journal of Community Service Learning*, *10*(3), 76–88. https://eric.ed.gov/?id=EJ905163

Shekhani, S., Iqbal, S., & Jafarey, A. (2021). Adapting the ethical review process for COVID-19 research: Reviewers' perspectives from Pakistan. *Eastern Mediterranean Health Journal, 27*(11), 1045–1051. https://doi.org/10.26719/emhj.21.053

Smith, L. T. (2021). *Decolonizing methodologies: Research and Indigenous peoples*. Bloomsbury Publishing. https://play.google.com/store/books/details?id=EwA1EAAAQBAJ

Sofaer, N., Jafarey, A., Lei, R. P., Zhang, X., & Wikler, D. (2007). Unconditional compensation: Reducing the costs of disagreement about compensation for research. *Eastern Mediterranean Health Journal = La Revue de Sante de La Mediterranee Orientale = Al-Majallah al-Sihhiyah Li-Sharq al-Mutawassit*, *13*(1), 7. https://www.researchgate.net/profile/Daniel-Wikler/publication/6289477_Unconditional_compensation_Reducing_the_costs_of_disagreement_about_compensation_for_research_subjects/links/549349a60cf22d7925da2e80/Unconditional-compensation-Reducing-the-costs-of-disagreement-about-compensation-for-research-subjects.pdf

Steen, M. (2013). Co-design as a process of joint inquiry and imagination. *Design Issues*, *29*(2), 16–28. https://doi.org/10.1162/desi_a_00207

Stilgoe, J., Lock, S. J., & Wilsdon, J. (2014). Why should we promote public engagement with science? *Public Understanding of Science*, *23*(1), 4–15. https://doi.org/10.1177/0963662513518154

Sutton-Brown, C. A. (2014). Photovoice: A methodological guide. *Photography and Culture*, *7*(2), 169–185. https://doi.org/10.2752/175145214X13999922103165

Takach, G. (2017). Arts-based research in the pedagogy of environmental communication. In T. Milstein, M. Pileggi, & E. L. Morgan (Eds.), *Environmental Communication Pedagogy and Practice* (pp. 102–111). Routledge. https://doi.org/10.4324/9781315562148–11

Talon, D. S. (2004). *Comics above ground: How sequential art affects mainstream media*. TwoMorrows Publishing. https://play.google.com/store/books/details?id=WHtNyhQ1-WAC

Tuck, E., & Yang, K. W. (Eds.). (2013). *Youth resistance research and theories of change*. Routledge. https://doi.org/10.4324/9780203585078

Tweddle, J. C., Robinson, L. D., Pocock, M. J. O., & Roy, H. E. (2012). *Guide to citizen science: developing, implementing and evaluating citizen science to study biodiversity and the environment in the UK*. NERC/Centre for Ecology & Hydrology. http://nora.nerc.ac.uk/id/eprint/20678/

Tyurina, A. (2022). Joining art and science to communicate water ecology. In C. Roulière & C. Egerer (Eds.), *Water lore: Practice, place and poetics* (p. 66). https://books.google.com/books?hl=en&lr=&id=Zw9qEAAAQBAJ&oi=fnd&pg=PT197&dq=Tyurina+A+(2022)+Joining+art+and+science+to+communicate+water+ecology+Water+Lore+Practice+Place+and+Poetics+66&ots=3Rt4OUws3R&sig=6DB_m5n6KPjtaplL1vOf-ilo2XE

UN. (2016). *Transforming our world: The 2030 agenda for sustainable development*. United Nations. https://stg-wedocs.unep.org/bitstream/handle/20.500.11822/11125/unepswiosm1inf7sdg.pdf?sequence=1

United States. National Commission for the Protection of Human Subjects of Biomedical and Behavioral Research. (1978). *The Belmont Report: Ethical principles and guidelines for the protection of human subjects of research: Appendix*. Department of Health, Education, and Welfare, National Commission for the Protection of Human Subjects of Biomedical and Behavioral Research. https://play.google.com/store/books/details?id=cAqtFohwYhIC

Valli, C. (2021). Participatory dissemination: bridging in-depth interviews, participation, and creative visual methods through Interview-Based Zine-Making (IBZM). *Fennia, 199*(1), 25–45. https://doi.org/10.11143/fennia.99197

Van Royen, K., Verstraeten, R., Andrade, S., Ochoa-Avilés, A., Donoso, S., Maes, L., & Kolsteren, P. (2015). Factors affecting physical activity in Ecuadorian adolescents: a focus group study. *Journal of Physical Activity & Health, 12*(3), 340–348. https://doi.org/10.1123/jpah.2013–0288

Vásquez-Guevara, D., Weiss, D., & White, J. M. (2022). Participatory co-design of science communication strategies for public engagement in the U.S. and Ecuador around health behavior change. *Research for All, 6*(1).

Vásquez-Guevara, D. H. (2019). *Science for all: Exploring science communication for public engagement in culturally diverse scenarios in the Americas* (Publication No. 13812526) [Doctoral dissertation, The University of New Mexico].

Verstraeten, R., Leroy, J. L., Pieniak, Z., Ochoa-Avilès, A., Holdsworth, M., Verbeke, W., Maes, L., & Kolsteren, P. (2016). Individual and environmental factors influencing adolescents' dietary behavior in low- and middle-income settings. *PloS One*, *11*(7), e0157744. https://doi.org/10.1371/journal.pone.0157744

Verstraeten, R., Van Royen, K., Ochoa-Avilés, A., Penafiel, D., Holdsworth, M., Donoso, S., Maes, L., & Kolsteren, P. (2014). A conceptual framework for healthy eating behavior in ecuadorian adolescents: A qualitative study. *PloS One*, *9*(1), e87183. https://doi.org/10.1371/journal.pone.0087183

Wallerstein, N., & Duran, B. (2006). Using community-based participatory research to address health disparities. *Health Promotion Practice*, *7*(3), 312–323. https://doi.org/10.1177/1524839906289376

Wallerstein, N., & Duran, B. (2010). Community-based participatory research contributions to intervention research: the intersection of science and practice to improve health equity. *American Journal of Public Health, 100*(1_suppl.), S40–6. https://doi.org/10.2105/AJPH.2009.184036

Wallerstein, N., Duran, B., Oetzel, J. G., & Minkler, M. (2017). *Community-based participatory research for health: Advancing social and health equity*. John Wiley & Sons. https://play.google.com/store/books/details?id=Gkk3DwAAQBAJ

Wang, C. C. (2006). Youth participation in Photovoice as a strategy for community change. *Journal of Community Practice*, *14*(1–2), 147–161. https://doi.org/10.1300/J125v14n01_09

White, J. (2013). Translating science, health and technology: Reporters as knowledge transfer intermediaries. *Online Journal of Communication and Media Technologies*, *3*(2), 227.

Whittington, E. (2019). Co-producing and navigating consent in participatory research with young people. *Journal of Children's Services*, *14*(3), 205–216. https://doi.org/10.1108/JCS-02-2019-0007

Chapter 4

Building Science Communication Strategies

New or Existing Initiatives

Denisse Vásquez-Guevara, Angélica Ochoa-Avilés, David Weiss, and Judith McIntosh White

After reviewing the available methodological paradigms of participatory methodologies, their data collection tools, and how these have been effectively used in science communication projects, we present two case studies developed by some of the authors of this book. In this endeavor, we aim to provide our readers with a full spectrum of how science communication strategies can be developed according to the needs and interests of audiences and scientists.

In Case Study 7, *A Todo Pulmón* (ATP), With Your Whole Lungs, we present a participatory research study we designed to develop informational and behavioral tools to support asthma self-management of children and teenagers and their caregivers in Ecuador, based on the reported barriers and limitations that young patients and their families face around this chronic illness.

Case Study 8 presents the results of a multiple-case study developed in two countries around two already-existing health behavior change programs, which promoted children's healthy nutrition and physical activity. This study explored how to develop engaging science communication in culturally diverse scenarios through a participatory evaluation approach that aimed to improve audience engagement.

Finally, we suggest three broad guidelines to plan science communication strategies and develop content and tools that use research findings to create collaboratively useful resources for audiences, with the goal of promoting audience engagement.

CASE STUDY 7: A TODO PULMÓN (ATP), *WITH ALL YOUR LUNGS: DEVELOPING* TOOLS FOR ASTHMA SELF-MANAGEMENT

In Ecuador, the pediatric asthma prevalence varies between 10 and 20 percent (Asher et al., 2021). Asthma prevalence is also linked to a high cost of medical treatments (Barnett & Nurmagambetov, 2011), low school-performance (Al-Durra et al., 2015), and social marginalization of young patients, which consequently brings emotional and psychological burdens to patients and their families, affecting their mental health (Rodríguez, Muñoz, & Muñoz, 2018; Lehrer et al., 1993). However, in the last two decades, asthma treatment has improved significantly in Latin American countries, such as Ecuador (Asher et al., 1995; Cabrera et al., 2021), making self-management a crucial factor for asthma treatments that include proper provider-patient communication (Ali & Young, 2022), environmental management guidelines of allergic triggers (Miles et al., 2017), medication use and adherence (Eakin & Rand, 2012), and healthy nutrition habits (Matos et al., 2011). In 2020, amid the COVID-19 pandemic the future of young patients was uncertain due to the lack of available studies and guidelines of the effects of COVID-19 on asthma patients.

For this reason, the Biosciences Department at University of Cuenca (UCuenca), in alliance with the Medicine Colleges of the University of Azuay (UDA), Central University of Ecuador (UCE), and International University of Ecuador (UIDE) developed the multidisciplinary research project *CEPRA XV 2021: Associated Factors with the therapeutic response in asthmatic children in the context of SARS-CoV-2*. The project was funded through a grant of the Ecuadorian Corporation for the Development of Investigation and Academica (CEDIA).

The research team combined biomedical studies and qualitative studies to explore and monitor the responses and symptoms of children and teenagers (5-to-13-year-olds) with asthma diagnosis and to explore patients' and caregivers' current knowledge and adherence to treatments regarding correct self-management guidelines. The project's goal was to create more understanding about the health consequences of COVID-19 and confinement in asthmatic children and teenagers, and the issues they and their caregivers/parents faced to follow self-management guidelines. The study also had a final health education goal: to create informational tools tailored to the patients' and caregivers' needs and knowledge.

During the first phase of the study, results evidenced that patients and caregivers were affected emotionally by the burden that the illness represented in their lives, misinformation about self-management techniques that involved natural treatments and practices (i.e., root syrups, infusions), which were not supported by scientific evidence and had, not only negative or non-positive effects on the patient's improvement but also replaced asthma medical treatments provided by doctors. Moreover, there were several misinformed health beliefs about the use of inhalers and corticoids (i.e., corticoids trigger child obesity) that represented patients' and caregivers' negative and misuse of medications that affected the overall asthma management. Additionally, parents and caregivers from the public health system, which were most of the participants—who were also from low-income and marginalized groups—had multiple questions about asthma that the specialists did not address at all or provided enough information.

This project required the design of an informational tool to support patients' self-management and knowledge about their asthma.

Methodological Approach

As we suggest in chapter 3, the first step is to use formative research to have enough understanding of the needs, interests, and current conditions of the target audience. Additionally, our research team carried out a focused literature review (FLR) (Mallett & Hagen-Zanker, 2018) on health communication studies that had effective outcomes to support patient asthma self-management. For this purpose, we used emerging data from the prior phases of the study to inform ourselves about the issues and concerns of young asthma patients and their caregivers, in order to determine how we will carry a co-design approach.

Co-design with Vulnerable Groups

In Case Study 7, the participants (children, teenagers, and their parents/caregivers) were considered a vulnerable population (Stowell et al., 2018). As asthma is a chronic illness (Riera et al., 2015), the level of misinformation among the study participants was high, there was a significant unattended emotional burden, and over half of the participants had a low-socioeconomic status. For these reasons, to approach this study, we used a qualitative approach with focus groups (Hamui-Sutton & Varela-Ruiz, 2013). To recruit participants, we used the main study's patient database, which included young patients and caregivers from the public and the private sectors. As a sampling strategy, we used non-probabilistic intentional sampling (Álvarez-Gayou, 2003) to select eighteen participant asthma patients between the ages of five and seventeen as well as their parents and caregivers divided into the following groups as we see in table 4.1.

Table 4.1: Asthma Patients by Age Group

Participant Groups	*Age group*	*Males*	*Females*	*Socioeconomic groups*
Children/Teenagers	7–9 years old	3	3	Medium and low-income
	10–12 years old	3	3	Medium and low-income
	10–12 years old	3	3	Medium and low-income
	Age of the patients	*Males*	*Females*	*Socioeconomic groups*
Parents or caregivers of the asthma patients	7–9 years old	3	3	Medium and low-income
	10–12 years old	3	3	Medium and low-income
	13–17 years old	3	3	Medium and low-income

Developed by authors.

Data Collection Tools

To develop the co-design process, we conducted two focus group sessions. During these sessions, we specifically consulted participants about their most frequent questions and doubts about their asthma self-management (communication with their medical specialists and informational needs about their asthma). Regarding the communication co-design, we drafted ideas for communication strategies by defining a name for the project, the communication conduits children and teenagers as well as their caregivers used more frequently, and their hobbies. Additionally, each focus group session had an asthma medical specialist to clarify any questions and/or misinformation that emerged (Koster et al., 2015).

To conduct the sessions, we used two semi-structured questionnaires to facilitate dialogue and participation in each focus group session (McIntosh & Morse, 2015).

Moreover, this study was conducted amid the sanitary emergency during the COVID-19 pandemic, with partial lockdowns, that limited in-person meetings, we carried the co-design sessions through the Zoom video conference platform.

Additionally, we used the expert panel method or Delphi Model (Varela Ruiz & Vives Varela, 2017) to validate the accuracy of the medical content and suggestions of all the communication strategy tactics to avoid misinformation. For this purpose, we invited two medical experts (one pediatric asthma specialist from the public health system and one from the private health system), and two health communication experts to evaluate and provide feedback about the strategy. In order to provide the experts with an online-based evaluation of the materials, we developed an electronic form, using evaluation guidelines, which was sent to experts asking them to provide us with a Linkert-scale evaluation and feedback comments (Palmieri, 2017).

Co-design Tools

In order to explore which messages, communication conduits, and hobbies young patients had, we used participatory diagrams (Kesby, 2000). Through the interactive-website platform *Mentimeter,* we digitized these diagrams, inviting people to vote and send their ideas through access links sent to them. The responses were presented to the focus group participants through Zoom's share screen feature to dialogue with the participants.

The semi-structured questionnaires used to conduct the focus group discussions explored patients' and caregivers' knowledge. They explored several health education communication strategies for asthma self-management based on: first patient-provider communication regarding the relationship patients have with their physicians, the clarity of the information, and the self-management practices for their asthma treatment (Ali & Young, 2022); and second, the tailored communication for health behavior change criteria explored the language tones, ideal communication conduits, preferred entertainment contents or activities (Noar et al., 2011).

With the suggestions and ideas provided by patients and their caregivers, the science communication team of UCuenca's Biosciences Department (two strategic communication researchers and practitioners, one copywriter, one social media manager, two video producers, and two graphic designers) developed the tactics drafted with the study participants for different formats that we present in the next section, results, and outcomes.

Results and Outcomes

As a result of the sessions, we collectively developed A Todo Pulmón. Young patients and their parents frequently commented on the burden of managing a chronic illness, and how much they needed to clarify information that helped them to better understand their self-management routines and adhere to their medications.

Children and parents gave the idea of the name and the logo for the health education tool, which was developed and refined by the graphic designers in the Biosciences Department, as illustrated in figure 4.1.

Core Self-management Informational Needs. Young patients and parents/caregivers identified six core areas of interest in asthma self-management. Frequently, patients had different misinformed beliefs about (1) what asthma is, (2) treatments, (3) healthy nutrition, (4) physical activity, (5) daily self-care, and (6) COVID-19 prevention.

A TODO PULMÓN
RESPIRA SANO

Figure 4.1. Logo of A Todo Pulmón, Breathe Healthy. ***Developed by Health, Nutrition, and Physical Activity Research Group at the University of Cuenca, Biosciences Department.***

We defined these as six broad topics and explored the subthemes regarding how they would like to be informed about each of these through specific communication formats, messages, and practical activities. As we also reviewed in chapter 3, co-design implies including people as active collaborators and co-researchers in the research process. For this reason, participants were invited to focus groups to give ideas about the ideal format in which health education tools must be developed.

Defining Messaging Content Strategies and Communication Conduits. Young patients and their caregivers/parents agreed that social media was a key communication conduit, especially through short videos using Instagram Reels, TikTok, and YouTube. Additionally, parents and young patients were invited to be part of videos and tutorials that demystified misinformation such as pets and asthma triggers, allergens at home, workout routines that can be done at home for asthmatics, and healthy recipes created by a food scientist of our research group for breakfast and lunch.

Moreover, the participants needed the tool to provide clear messages that required researchers and medical specialists to be the main spokespeople for important messages, such as nutrition, treatment, and asthma education. Participants reinforced our reflections on chapters 1 and 3. As we learned before, people want to meet experts and learn from them. In order for this to happen, experts must be visible and present in educational tools to reinforce audiences' trust by using a clear, approachable, and non-technical language, but above all, these tools must give audiences clear guidelines to manage asthma.

For this reason, we created six videos of one to five minutes to instruct audiences about asthma, the correct way to use inhalers must, healthy nutrition, and its relation to creating a strong immune system. The videos were developed for the social media platforms Instagram Reels and TikTok.

Age-based Interactive Strategies. Children and Teenagers had different interests. While children liked arts-based activities such as coloring books, and crafts, teenagers frequently enjoyed reading anime comic strips by independent graphic artists online as a hobby in their spare time.

For five-to-eight-year-old children, we created an activity coloring book named *Asmaquipedia,* downloadable and printable format, which includes labyrinths, coloring pages, alphabet soups, crosswords, and find the seven differences games (figure 4.2). Additionally, since young audiences had several difficulties identifying allergens and asthma attack triggers in their homes, we created a board game of a DIY home (figure 4.3) to identify key practices such as ventilating rooms every day, eating healthy snacks, and keeping pets away from bedrooms. Children could download the game from our social media and website and print a coloring version or a black-and-white version to assemble and play with it, thus learning more about healthy habits.

For teenagers, we decided to create three main anime characters (figure 4.4) and develop a story contest where patients were invited to write a motivational story about how to overcome asthma. The contest would be promoted through social media to invite teenagers to participate. A jury of the co-design team (researchers and patients) would pre-select the best stories, and these would be presented to audiences from the different university programs on social media platforms for them to vote on the best story. The winners will have their story fully animated in audio and video with the ATP characters and will have a special prize from the participant universities.

However, all the young patients from different age groups used their smartphones (with limited access to social media according to their ages) to set alarm reminders for their medication times and the DM messaging platform

Figure 4.2. Asmaquipedia activity pages that were co-created with participants. ***Developed by Health, Nutrition, and Physical Activity Research Group at the University of Cuenca, Biosciences Department.***

WhatsApp to communicate with their peers, parents, and even their physicians of the private sector (figure 4.5). For this reason, we created sticker messages with short motivations and emojis (*you can do it, only breathe, breathe deeply-it's over, you are one step away from breathing healthy*).

COVID-19 Guidelines. Patients and their caregivers agreed that they were worried about their health vulnerabilities because of the easing of lockdown measures and restrictions to contain the COVID-19 spread and contagion. Specifically, one main concern was the school's in-person attendance. For this reason, the team's medical researchers, patients, and their caregivers developed an infographic with recommended guidelines for asthma patients returning to school. (See figure 4.6.) Patients identified key points, such as the prolonged use of masks, social distancing, surface disinfection, and personal hygiene.

Expert Review and Refining Science Communication

After the first focus group, all the ATP health education tools were sent to the experts' panel to have their medical content and educational communication

Figure 4.3. The DIY board game of ATP to learn about allergens and healthy habits for asthma patients. ***Developed by Health, Nutrition, and Physical Activity Research Group at the University of Cuenca, Biosciences Department.***

Figure 4.4. Ari, Dr. Nuna, and Samai, the ATP's anime characters. *Developed by Health, Nutrition, and Physical Activity Research Group at the University of Cuenca, Biosciences Department.*

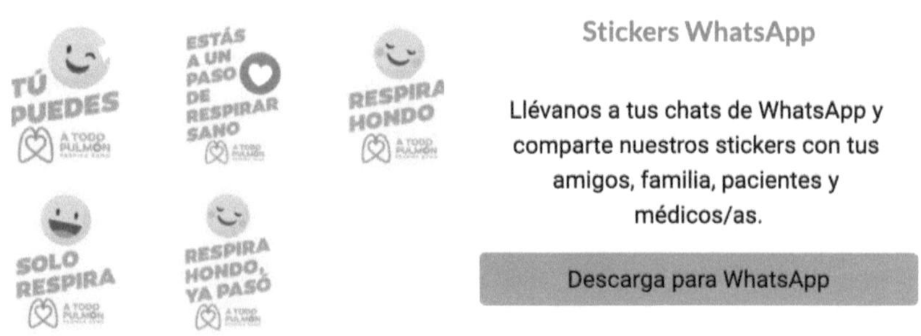

Figure 4.5. WhatsApp stickers for ATP. *Developed by Health, Nutrition, and Physical Activity Research Group at the University of Cuenca, Biosciences Department. These stickers can be downloaded at: https://sticker.ly/s/BD2TRE*

Figure 4.6. Infographic: Are you prepared to go back to school? ***Developed by Health, Nutrition, and Physical Activity Research Group at the University of Cuenca, Biosciences Department.***

strategy content reviewed. Asthma specialists provided feedback regarding specific iconography, caption texts that should be included in the videos, and activities to strengthen key information. For example, medical and communication experts agreed that the medical recommendations should have clear disclaimers regarding strictly following specialists' guidelines.

After receiving the comments and feedback suggested experts panel, the health education tools, and their activities were revised and adapted to be presented again to the study participants; they were invited to use them and to give comments to refine any further details based on their feedback.

This case illustrates a scenario where health education tools were created from zero with the collaboration of patients, experts, and science communication researchers and practitioners, as part of a multidisciplinary project. ATP's health education tools will be used in the future for a large-scale medical-intervention research project that combines biomedical research and resources for young asthma patients in the public and private health system in Ecuador.

Key Considerations from Health Communication and Science Communication to Build Effective Strategies

To create the health education tools, we developed a qualitative and participatory approach to establish a co-design process that united researchers, medical specialists, asthma patients, and their parents or caregivers. As science communication frequently develops research and practice about topics of different disciplines (i.e., health, environmental sciences, gender studies) it is crucial to inform the development process with strong scientific evidence in terms of content, but also about models that can support creating an effective communication approach that contributes useful information.

For this multidisciplinary project, we used several guidelines, models, and theories from health communication and science communication. From health communication, we used health behavior change of the socio-cognitive theory model (SCT) (Bandura, 2004). SCT provided us a framework to understand the interaction of environmental criteria such as how peers at school and family influenced asthma patients' behavior to adopt or not self-management guidelines suggested by their audiences. Additionally, we also used the tailored messaging notions of health communication to draft several messages for communication tools and content according to the audiences' informational needs and (Noar et al., 2011; Rimer & Kreuter, 2006).

From science communication, we used several of the sixteen considerations for its effectiveness, developed by Steven Cooke, Austin Gallagher, Natalie Sopinka, Vivian Nguyen, Rachel Skubel, Neil Hammerschlag, Sarah Boon, Nathan Young, and Andy Danylchuk (2017). Specifically, we used the following guidelines and adapted them to the co-design process of this specific project (Cooke et al., 2017):

1. *Define What Science Communication Means to You and Your Research.* As a starting point, this guideline centers on situating the main goals to develop a science communication strategy from the researcher's perspective and defining the target audience. Science communication can be useful to "increase the spread of knowledge and (or) be a driving force in changing policy or decision-making" (Cooke et al., 2017 p. 235). In this project, we had the clear goal of creating health education tools to support young patients' asthma self-management. Therefore, we focused on identifying core situations where asthma self-management guidelines could be clearly shown to asthma patients and their caregivers, for example: correct medication use, healthy nutrition, physical activity, and allergen management at home.
2. *Know—and Listen to—Your Target Audience.* This guideline refers to properly knowing and understanding the target audiences regarding their

cultural norms, worldview, personality, and demographic characteristics through studies and observation to develop a suitable and feasible strategy to approach them. Consequently, first, we used the outcomes of a prior study—of the same project—as formative research to understand better asthma patients' needs and hurdles around self-managing their treatments. Second, we invited them to participate voluntarily in co-designing health education tools tailored to their needs. For this reason, we separated sections of each focus group to center dialogue around their experiences with physicians and understanding of their daily routines as asthma patients and to discuss frequent questions.

3. *Consider a Diverse but Coordinated Communication Portfolio.* Cooke and colleagues suggest combining scholarly dissemination activities (i.e., publishing papers, developing scientific posters, or attending scientific events) with media strategies that focus on communication conduits target audiences frequently use, such as social media, press interviews, or discussion sessions with community groups. For ATP we developed social media-based science communication strategies according to the communication conduits preferred. Consequently, we focused on creating specific infographics, art-based games, and video content using Instagram and TikTok.
4. *Be Creative When You Communicate.* This guideline challenged our science communication team to explore our creativity. Cooke et al. invited us to reflect on developing multidisciplinary strategies that combined art, journalism, gaming strategies, and enjoyable activities through the lens of our young audiences' main needs: motivations to embrace the challenge of self-managing their asthma.

In our co-design process, we were aware of the emotional burden that the illness represented to children, teenagers, and their parents/caregivers; and the frequent misinformation about crucial practices for correct asthma self-management. For this reason, we decided to explore with them all the activities that were their favorite hobbies and create health education communication pieces with them. As a result, we obtained DIY activities, anime comic characters, and video tutorials for healthy recipes and workout routines that asthma patients can do at home.

Health communication and science communication theoretical models were combined to develop an accurate strategy that was tailored to the target audiences' needs, interests, and concerns. The preliminary revisions in the second focus group to pretest the materials with asthma patients and their caregivers presented positive results and responses from them. As one in this study, we suggest that once materials, resources, or strategies are ready, these must be revised through pre-testing strategies to refine details before large-scale dissemination plans or as part of interventions.

Science communication, as we have seen in this case as well as, in the rest of the book, is a multidisciplinary effort in research and practice. In order to produce quality materials, and activities, having enough resources and specialized professionals to design, implement and execute strategies, resources, or any initiative to communicate science effectively to the broader public or non-experts is required. However, it is an effort that offers several opportunities to build sustainable relationships with society, and communities by offering them scientific content that not only informs and educates people but offers tools and solutions tailored to specific problems.

EVALUATING AND RETOOLING SCIENCE COMMUNICATION OF EXISTING PROJECTS

As we mentioned before, science communication can be adapted to different scenarios, such as evaluating existing projects that present the results of interventions, research, or science education campaigns.

In the following case, we review how to evaluate and retool already-existing programs to develop science communication strategies that promote audience engagement. Moreover, we explore how to involve former audiences as co-researchers that inform the process of participatory methodologies through their local knowledge and lived experiences.

The outcomes evidence the versatility of PAR for science communication research and practice.

Case Study 8: Science for All, Exploring Science Communication for Public Engagement around Health Promotion Programs

The present study combines qualitative and participatory action research methods to develop a co-designed project with public health, nutrition, and sports science researchers to recruit young audiences from Albuquerque, New Mexico, in the United States, with the *Eat Smart to Play Hard* (ESPH) Health Promotion program, and ACTIVITAL Cuenca, Ecuador.

Eat Smart to Play Hard

Eat Smart to Play Hard (ESPH) is a health promotion program created and implemented by the University of New Mexico's Prevention Research Center from the UNM Health Sciences Center. ESPH is a program with six years of experience in reducing childhood obesity and preventing chronic diseases in families, schools, and communities across the state of New Mexico. ESPH applies a social marketing approach (Grier & Bryant, 2005) on a series of interventions at elementary public and private schools and in households to motivate children to adopt healthy nutrition and physical activities (Applegate, 2001). Moreover, ESPH uses a habit-building approach through a six-week program using their *Fun Book,* a booklet that suggests daily workouts and healthy recipes to prepare meals at home and snacks for school, parent newsletters, and school-based activities.

ACTIVITAL

ACTIVITAL is a health promotion program developed by University of Cuenca's (UCuenca) Biosciences Department at the University of Cuenca in Ecuador and the VLIR Program Cooperation Alliance-Belgium. ACTIVITAL was implemented in 2016 to address the broader issue of children's unbalanced nutrition in Ecuador, characterized by disproportionate food intake, driven by cultural perceptions, parents' and caregivers' unhealthy cooking habits, which dramatically influence children's and teenagers' health conditions. The ACTIVITAL program was developed through a socio-ecological approach in order to educate children, teenagers, and their families about healthy nutrition, using school interventions that included group games, workshops, medical controls, and the collaborative development of a healthy-eating recipe book (Ochoa-Avilés et al., 2017).

This study developed a qualitative multiple-case study to create engaging science communication for both programs by organizing four participatory workshops to evaluate the experiences of children (ESPH) and teenagers (ACTIVITAL) in two countries with the researchers of each program. To conduct PAR workshop sessions, we used PAR data collection tools, such as participatory diagrams, asset mapping, and zines.

The PAR sessions allowed us to identify strategies of each program that needed to be readapted or changed. The PAR exploration of ESPH demonstrated that the program needed to understand better their audiences' context dictated by their family lifestyles and household income, which made it difficult for participants to afford vegetables and proteins in New Mexico and to do physical activity in parks and public playgrounds due to street unsafety.

On the other hand, ACTIVITAL needed to update its school-based intervention activities, and include online and offline informational resources, such as incorporating social media, a healthy habit tracker app, and public events for the broader community.

As we described in the Science for All study, ESPH and ACTIVITAL wanted to review their science communication strategies in order to improve

Finally, it is important to mention that all the co-researchers received a $40 USD compensation in each country, as well as healthy snacks and co-design materials for developing science communication strategies.

Exploration of Existing Strategies and Co-design of Alternative Strategies

As we mentioned before, we conducted four PAR workshops for each program. To conduct these PAR sessions, we developed prompt questions that guided the work session discussions. Prompt questions are open-ended questions that have the purpose of focusing on specific issues and inviting people to dialogue and explore their ideas (see table 4.2). Researchers use prompt questions are used by researchers to collect information about their perspectives and lived experiences and initiate two-way dialogic sessions (Eggly et al., 2013).

Moreover, each workshop had a specific goal in the project to develop an effective exploration of science communication strategies that promoted audience engagement.

Workshop 1. The first discussion centered on assessing the communicational pieces of each case study. As a data collection tool, PAR diagramming (Kesby, 2000) was used to organize and select emergent themes for discussion and to connect causes and consequences with the alternatives proposed as solutions for emergent problems and with issues that required improvement. As we show in figure 4.7, we were able to connect the assessment of each communicational piece of ESPH and ACTIVITAL, explain the causes of the issues of difficulties people faced difficulties with each strategy, and propose solutions.

Workshop 2. In this session, we focused on exploring more in-depth how to improve audience engagement in the programs. In this session, we used PAR diagrams to develop the alternative strategies and how these should function as well as, the asset mapping (Canada Excellence Research Chair in Migration & Integration, 2021) to allocate places, and resources that we could incorporate as part of the co-designed strategies.

For this purpose, we focused on co-designing strategies that strengthened the positive tactics of each program, but from the audiences' perspective, it required specific adaptations. For example, ESPH's *Fun Book* motivated children to cook healthy meals, but sometimes their families' low incomes limited the affordability of vegetables and proteins in New Mexico.

Consequently, in this second session, we explored how to support audiences better. Again, we returned to the first PAR diagram and discovered that the prizes ESPH offered were attractive to children (e.g., water bottles, medals, playballs); however, some of these could be replaced by more useful resources, such as gift cards from grocery stores that allowed them to afford

Table 4.2. Prompt Questions

Workshop 1	• *What do you think about the activities that ACTIVITAL/ESTPH have developed to communicate their program? How well do you think they are working?* • *Which activities of ACTIVITAL/ESTPH do you think are most effective? Why?*
Workshop 2	• *If you could design an activity for communicating ACTIVITAL/ESTPH to teenagers like you, what would you do/suggest? Think about something you would like to see or do and other activities that teens like you might enjoy and engage in.* • *Where can (or where do) teenagers find information about healthy habits?* • *Which venues are available for children and parents to do physical activity and play in this city?*
Workshop 3	• *What activities will you develop to create engagement with your audience around issues of healthy eating/activities?* • *What would you add/change to ESTPH/ACTIVITAL?*
Workshop 4	• *How can we explore further engagement opportunities for ESTPH/ACTIVITAL?*

Developed by Vásquez-Guevara (2019).

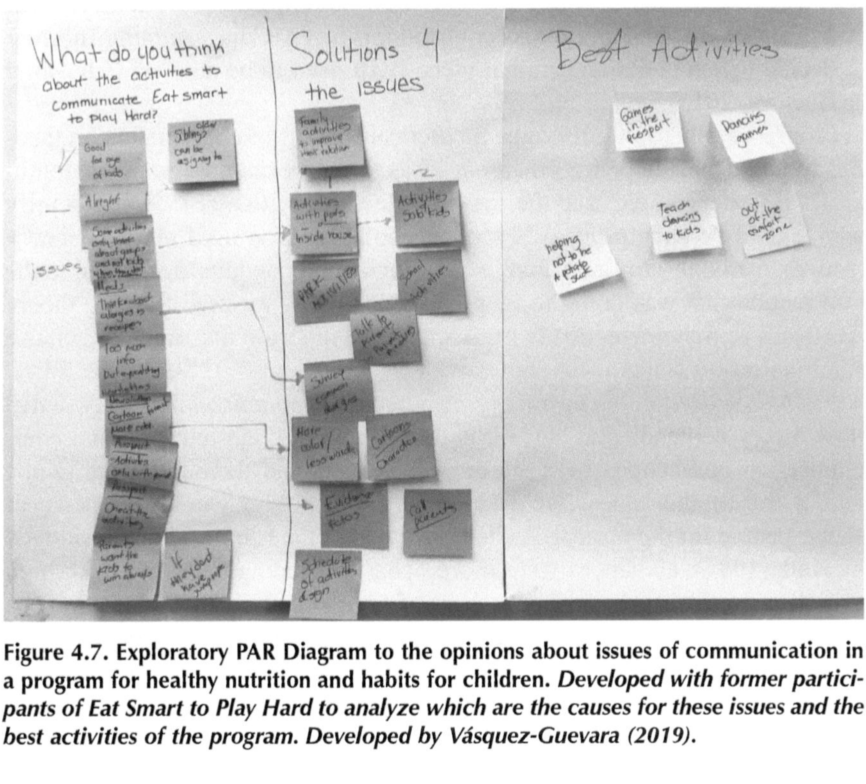

Figure 4.7. Exploratory PAR Diagram to the opinions about issues of communication in a program for healthy nutrition and habits for children. ***Developed with former participants of Eat Smart to Play Hard to analyze which are the causes for these issues and the best activities of the program. Developed by Vásquez-Guevara (2019).***

more healthy food alternatives. In this case, asset mapping was useful to determine which were the most frequented grocery stores to get gift cards, which represented interesting prizes to audiences.

In the case of ACTIVITAL, PAR diagrams and asset maps revealed that the program needed to offer not only school-based interventions, such as workshops but to also innovate and use an online and offline strategy approach. PAR diagrams pointed out that the school-based intervention was very useful, although sometimes the co-researchers could not recall specific data about food portions and forgot key information. For this reason, they suggested creating social media content on Facebook, Instagram, and YouTube with tutorials and tips that reminded them of what was presented in-school interventions.

Regarding offline strategies, co-researchers also suggested doing open events for the community at riverside parks in Cuenca on weekends, so families could attend together to do workout routines, cooking workshops, and dialogue events to ask experts and researchers about healthy nutrition.

Workshop 3. This session focused on defining the messaging strategies for each program. For this purpose, we needed to explore the existing texts of each program's communication strategies and suggest any changes or observations. In this session, we also used participatory diagrams to rank and evaluate the attribution and contribution of PAR diagramming in order to decide which communicational pieces will need to be adapted (Chevalier & Buckles, 2013).

From this point on we used the theoretical support of communication theories to guide and shape the emergent ideas to retool each program's communication strategies. We used the two-step flow theory (Katz, 1957) to identify who should be each program's spokespeople; we also used medium theory (Meyrowitz, 2009) to find suitable communication conduits, according to the information that was going to be presented; finally, we used framing theory (Brossard & Scheufele, 2013; Entman, 1993) to shape messages of science communication content.

Finding Spokespeople for Engaging Science Communication. The two-step flow theory contends that content is transmitted through interpersonal communication interactions between opinion leaders and their followers (Katz, 1957). So, in this study, we wanted to explore who would be the ideal spokespeople for the project. As demonstrated in the Elmira Study developed by Katz (1957), audiences were more open to listening to people with high expertise. In this study, co-researchers of both countries and programs conclude that the best spokespeople for health information should be scientists themselves, especially when the programs needed to present key health information about nutrition and physical activity and their effects on audiences.

Additionally, in the specific case of ESPH children suggested creating the *Eat Smart to Play Hard Squad* (see figure 4.8), in order to offer audiences a more personal experience with characters they could relate to. The ESPH Squad united different characters that reflected the program and all the people involved in the effort to motivate others to embrace a healthy lifestyle. Among the characters and their roles this was proposed:

- *Scientists*. Researchers of the ESTPH program should be part of ESPH communication by using games and experiments to explain the science behind food and physical activities.
- *Teen Health Educators*. This group of opinion leaders should focus on the fun of physical activity and healthy eating. Their goal would be to design games and provide support by talking with children about how to incorporate healthy habits into their daily routines. Moreover, the teens selected for this role should be role models for children by promoting a healthy lifestyle. Teenagers would be invited as volunteers to educate children through games and activities that would be enjoyable and support the ESPH claims.
- *Children*. The former participants of the program who were successful in following ESPH and made changes in their lives would be invited to be part of the Squad. They would tell their stories to other children about how they and their families learned and made improvements in their health.
- *A pet*. In all four sessions, dogs and cats were always part of the co-researcher's conversation. Most of the children had a pet at home. Pets were described as "unconditional companion," which is always there for the children and are generous with their love. The pet could be in the company of the children during the campaign as well as serve as

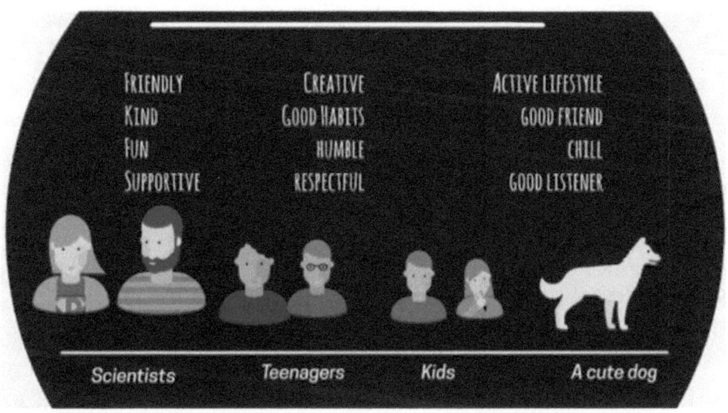

Figure 4.8. The ESPH squad. ***Developed by Vásquez-Guevara (2019).***

a playmate during physical activity. The children could play throw and catch with their dog, walk their dog, and do pet yoga.

In ACTIVITAL, the Ecuadorian co-researchers also concluded that scientists must be visible and present health recommendations to parents and teenagers. In this case, they wanted to have dialogic spaces with physicians and researchers, such as webinars or workshops on social media (e.g., Instagram Live events), and in-person meetings they can ask specific questions about their health. Additionally, co-researchers suggested that the program includes its former participants, who could provide testimonials about their experiences with embracing healthy habits and their direct benefits in their lives.

Suitable Conduits for Science Communication. As we mentioned before, we used medium theory as a guide to reflect on how to find suitable communication conduits for our audiences and the specific content of the programs. The medium theory provided useful theoretical guidelines to reflect on how and which communication conduits, analyzed in terms of their "biases" (that is, their characteristics and features), might or might not be effective to communicate science content for each case study (Meyrowitz, 2009). Moreover, medium theory along with framing theory was useful to choose the best format (i.e., printed media, social media, videos, games, and events) to engage audiences around key health practices.

Here we found that in the current scenario, traditional media and online media needed to be combined in order to be effective. For example, printed media and press events needed to be integrated with online conduits such as social media and apps in order to reach broader audiences. Social media were the most suggested conduit due to the structural characteristics that each platform offers to present different communication formats (Collins et al., 2016). YouTube, Instagram, and Facebook were the most popular choice to present videos of different formats and lengths, live events, and infographics.

In both programs, we found that in order to offer effective science communication strategies, we must develop tailored messaging strategies that attended to audiences's interests and needs (Noar et al., 2011; Rimer & Kreuter, 2006). Regarding the language tone, ESPH co-researchers suggested focusing on interpersonal communication that flows from the ESPH Squad to the audiences with a "fun, happy and healthy tone" as a friend speaking to them and suggesting how to make their habits better.

On the other hand, in ACTIVITAL, co-researchers suggested using local teenagers' slang to talk to their peers and use positive messages. For message framing, the use of testimonials, and storytelling techniques to present positive real-life experiences ACTIVITAL's former participants. For example, former participants could appear on a video tutorial preparing a healthy recipe in their home kitchens that they learned in the program. This strategy is also known in health communication as role modeling tactics derived from social learning (Ottoni-Wilhelm et al., 2014), from which young people learn specific behaviors through the influence of their social network, parents, and peers.

Specifically, framing theory is crucial for science communication, as it appeals to specific mental schemas that people can identify with and understand to make meaning about complex data. As science journalists write a piece that involves science, they might use comparisons to daily experiences in order to support the audience's understanding (Listerman, 2010).

Workshop 4. In this last session, we organized the overall strategies created for each program. For this purpose, Zines (Guzzetti & Gamboa, 2004) and PAR diagrams (Kesby, 2000) were used. Through PAR, we established a list of shared interests and an inventory of people, possible stakeholders, and participants' and researchers' skills and interests that could be useful for further collaborative initiatives. This step served as a list of potential collaborators that could support the program's dissemination and alliances that could be made with schools, and grocery stores to execute the new co-designed strategies.

On the other hand, zines were conducted to explore how to structure a campaign book for ESPH and ACTIVITAL that organized the science communication strategies for each program. The zines were handmade and combined cutouts and texts to present ideas for artwork, messages, and communication conduits for each tactic in order to have a well-developed and organized strategy to motivate audiences' engagement.

General Guidelines for Retooling Science Communication Initiatives in Research and Practice

Developing assessments of already-existing initiatives implies several procedures and steps. In this section, we will summarize our recommendations to conduct science communication assessment and facilitate the development of new proposals to improve strategies and tactics in a participatory methodological framework.

Assessment

In order to assess existing projects, we need to collect all the information, strategic planning, tactics, communication pieces, and available records of implementation or if available evaluation reports. Here, it is crucial to run a careful review to: first, understand the overall functioning of the project, and second, the goals and specific strategies and tactics that were implemented to achieve these goals.

We suggest that the assessment phase includes interviewing the project developers and implementation teams. In these groups, we can find research scientists, communication practitioners, and project staff who have first-hand experiences and direct contact with target audiences and the involved stakeholders. For this reason, their experience is necessary to identify key points of improvement that need to be adapted or changed. For example, in the Science for All study presented in this chapter, even when children and teenagers were the main target audiences, their school, and high school teachers were able to provide useful information about the school-based interventions carried out by each health promotion program. Here, teachers mentioned relevant aspects, such as children's and teenagers' prior knowledge about nutrition, and how the programs impacted their habits, as evidenced in the change in food choices for lunch.

Therefore, the science communication team needs to keep a record of this information in order to find useful guides to carry out a participatory evaluation process.

Retooling or Adaptation Processes

Once the complete background information is complete, we need to identify which science communication goals are needed for the specific project. In our own experience, retooling processes might imply adapting, changing, or developing new strategies to replace specific science communication activities or tactics in order to better serve the projects' goals.

Some of the latest studies that explore common science communication goals identify that projects frequently seek to train researchers on building communication skills knowledge with audiences, fostering excitement, building trust, and reframing issues from audiences' perspectives (Besley et al., 2016). Moreover, science communication, in the last decades, has focused on making new scientific advances visible to the public to make informed decisions about different issues concerning their lives (Weingart & Joubert, 2019), and to support behavior change processes by offering audiences to make better-informed choices (Fischhoff, 2019). After identifying the goals a project has around an adaptation or retooling process we need to move on to planning the different requirements and resources that are necessary to carry out the process.

Regarding resources, after identifying the goals and having enough background information about the project, the first step of the retooling process must be to reflect about the material and human resources needed to execute the project. On one hand, the material resources refer to the project's available funding for its development and further implementation. On the other hand, regarding human resources, we refer to the experts or practitioners and the specific skills that are required to carry out such adaptations. Science communication projects frequently require communication practitioners who create communication pieces, (e.g., graphic designers, copywriters, and audiovisual experts), science communication or public information practitioners who interpret scientific information and develop a communication strategy; and implementation teams who must be professionals with strong interpersonal communication skills and professional knowledge. We suggest that this last sub-team is composed of science communication practitioners and expert researchers, which can be the authors of their very own research studies. As we have argued throughout this book, a common issue in science communication is misinformation (Friedman et al., 1999), and scientists' active involvement in science communication is crucial to motivate audiences' engagement (Vásquez-Guevara, 2019, 2021). In science communication, evaluation processes must conversely support the roles of practitioners and researchers to connect with project goals and each other's work by focusing on identifying the results that matter to key audiences, and from this point structuring different strategies and their tactics (Fischhoff, 2019).

A second step must choose an evaluation or assessment methodology to explore which specific sections of a project must be improved. Since we are referring to projects of science communication for public engagement in this book, we have applied PAR methods in different scenarios that invite audience collaboration along with scientific endeavors.

Other methodologies to evaluate the effectiveness of science communication initiatives include quantitative and qualitative measures based on several baseline criteria of specific science and technology programs established by countries or scientific organizations. The report developed by Toss Gascoigne and Jenni Metcalfe (2001) revealed the need to consider evaluation processes for the public communication of science and technology. These evaluation processes need to be built in from the start of a program, during the design stages, and be later compared with baseline data of the population's conditions before the program's implementation. We can agree to this suggested framework, as a comprehensive approach to science communication projects' performance evaluation. However, we agree with the report in order to suggest that PAR methods can be used to adapt or make changes while projects are implemented, are underway, or are waiting until the end of the implementation to suggest changes informed by the results and/or audiences' opinions (Gascoigne & Metcalfe, 2001).

Retooling the science communication of existent programs requires including the audience's opinions and lived experiences, especially, if public engagement is the main goal. Here, we advocate the use of PAR methodologies and tools to invite audiences to participate along with the researcher in exploring how audiences perceived the initiative's contents and activities were perceived. Moreover, PAR methodologies offer additional opportunities, as it invites audiences to suggest from their own perspective how science communication strategies could be in order to be more interesting to similar audiences and consequently develop a more engaging approach.

As we presented in chapter 3, there are multiple PAR genres that can be useful according to the audiences' needs and interests. Additionally, the different PAR tools for data collection and facilitation are useful to co-create different strategies that unite scientists and people's collaborative work of scientists and people.

Finally, retooling processes must develop a repository of the outcomes of collaborative work and how the new strategies reinforce or replace former tactics. Specifically, we suggest developing reports that suggest guidelines for implementation and further evaluation.

In chapter 5, we present several broader guidelines to develop science communication research and practical initiatives based on several theories and models, which collect our experiences and relevant literature in order to orient researchers and practitioners on how to develop engaging science communication.

BIBLIOGRAPHY

Al-Durra, M., Torio, M.-B., & Cafazzo, J. A. (2015). The use of behavior change theory in Internet-based asthma self-management interventions: A systematic review. *Journal of Medical Internet Research*, *17*(4), e89. https://doi.org/10.2196/jmir.4110

Ali, A. M., & Young, H. N. (2022). Relationships between key functions of patient-provider communication, trust, and motivation across white, African American, and Hispanic/Latino patients with asthma. *Health Communication*, *37*(4), 450–456. https://doi.org/10.1080/10410236.2020.1848758

Álvarez-Gayou, J. L. (2003). *Cómo hacer investigación cualitativa. Fundamentos y metodología*. [How to do cualitative research: foundations and methodology]. Paidós Mexicana.

Applegate, L. (2001). *Eat smart, play hard.* Rodale Press. https://hsc.unm.edu/medicine/departments/pediatrics/divisions/pps/initiatives/esph/

Asher, M. I., Keil, U., Anderson, H. R., Beasley, R., Crane, J., Martinez, F., Mitchell, E. A., Pearce, N., Sibbald, B., & Stewart, A. W. (1995). International Study of Asthma and Allergies in Childhood (ISAAC): Rationale and methods. *The European Respiratory Journal: Official Journal of the European Society for Clinical Respiratory Physiology*, *8*(3), 483–491. https://www.ncbi.nlm.nih.gov/pubmed/7789502

Asher, M. I., Rutter, C. E., Bissell, K., Chiang, C.-Y., El Sony, A., Ellwood, E., Ellwood, P., García-Marcos, L., Marks, G. B., Morales, E., Mortimer, K., Pérez-Fernández, V., Robertson, S., Silverwood, R. J., Strachan, D. P., Pearce, N., &

Shah, J. (2021). Worldwide trends in the burden of asthma symptoms in school-aged children: Global Asthma Network Phase I cross-sectional study. *The Lancet*, *398*(10311), 1569–1580.

Bandura, A. (2004). Health promotion by social cognitive means. *Health Education & Behavior: The Official Publication of the Society for Public Health Education*, *31*(2), 143–164. https://doi.org/10.1177/1090198104263660

Barnett, S. B. L., & Nurmagambetov, T. A. (2011). Costs of asthma in the United States: 2002-2007. *Journal of Allergy and Clinical Immunology*, *127*(1), 145–152.

Besley, J. C., Dudo, A. D., Yuan, S., & Abi Ghannam, N. (2016). Qualitative interviews with science communication trainers about communication objectives and goals. *Science Communication*, *38*(3), 356–381. https://doi.org/10.1177/1075547016645640

Brossard, D., & Scheufele, D. A. (2013). Science, new media, and the public. *Science*, *339*(6115), 40–41. https://doi.org/10.1126/science.1232329

Cabrera, A., Picado, C., Rodriguez, A., & Garcia-Marcos, L. (2021). Asthma, rhinitis and eczema symptoms in Quito, Ecuador: A comparative cross-sectional study 16 years after ISAAC. *BMJ Open Respiratory Research*, *8*(1), e001004. https://doi.org/10.1136/bmjresp-2021–001004

Canada Excellence Research Chair in Migration & Integration. (2021). *Community asset mapping*. Toronto Metropolitan University. https://doi.org/10.32920/17041241.v1

Chevalier, J., & Buckles, D. (2013). *Participatory action research: Theory and methods for engaged inquiry*. Routledge. https://doi.org/10.4324/9780203107386

Collins, K., Shiffman, D., & Rock, J. (2016). How are scientists using social media in the workplace? *PloS One*, *11*(10), e0162680. https://doi.org/10.1371/journal.pone.0162680

Cooke, S. J., Gallagher, A. J., Sopinka, N. M., Nguyen, V. M., Skubel, R. A., Hammerschlag, N., Boon, S., Young, N., & Danylchuk, A. J. (2017). Considerations for effective science communication. *Facets*, *2*(1), 233–248. https://doi.org/10.1139/facets-2016–0055

Eakin, M. N., & Rand, C. S. (2012). Improving patient adherence with asthma self-management practices: What works? *Annals of Allergy, Asthma & Immunology: Official Publication of the American College of Allergy, Asthma, & Immunology*, *109*(2), 90–92. https://doi.org/10.1016/j.anai.2012.06.009

Eggly, S., Tkatch, R., Penner, L. A., Mabunda, L., Hudson, J., Chapman, R., Griggs, J. J., Brown, R., & Albrecht, T. (2013). Development of a question prompt list as a communication intervention to reduce racial disparities in cancer treatment. *Journal of Cancer Education: The Official Journal of the American Association for Cancer Education*, *28*(2), 282–289. https://doi.org/10.1007/s13187-013-0456-2

Entman, R. M. (1993). Framing: Toward clarification of a fractured paradigm. *The Journal of Communication*, *43*(4), 51–58. https://doi.org/10.1111/j.1460–2466.1993.tb01304.x

Fischhoff, B. (2019). Evaluating science communication. *Proceedings of the National Academy of Sciences of the United States of America*, *116*(16), 7670–7675. https://doi.org/10.1073/pnas.1805863115

Friedman, S. M., Dunwoody, S., & Rogers, C. L. (1999). *Communicating uncertainty: Media coverage of new and controversial science*. Routledge.

Gascoigne, T., & Metcalfe, J. (2001). Report: The evaluation of national programs of science awareness. *Science Communication*, *23*(1), 66–76. https://doi.org/10.1177/1075547001023001007

Grier, S., & Bryant, C. A. (2005). Social marketing in public health. *Annual Review of Public Health*, *26*(1), 319–339. https://doi.org/10.1146/annurev.publhealth.26.021304.144610

Guzzetti, B. J., & Gamboa, M. (2004). Zines for social justice: Adolescent girls writing on their own. *Reading Research Quarterly*, *39*(4), 408–436. https://doi.org/10.1598/rrq.39.4.4

Hamui-Sutton, A., & Varela-Ruiz, M. (2013). La técnica de grupos focales. *Investigación en educación médica*, *2*(5), 55–60. https://doi.org/10.1016/s2007-5057(13)72683–8

Katz, E. (1957). The two-step flow of communication: An up-to-date report on an hypothesis. *Public Opinion Quarterly*, *21*(1), 61. https://doi.org/10.1086/266687

Kesby, M. (2000). Participatory diagramming: deploying qualitative methods through an action research epistemology. *Area*, *32*(4), 423–435. https://doi.org/10.1111/j.1475–4762.2000.tb00158.x

Koster, E. S., Philbert, D., de Vries, T. W., van Dijk, L., & Bouvy, M. L. (2015). "I just forget to take it": Asthma self-management needs and preferences in adolescents. *The Journal of Asthma: Official Journal of the Association for the Care of Asthma*, *52*(8), 831–837. https://doi.org/10.3109/02770903.2015.1020388

Lehrer, P. M., Isenberg, S., & Hochron, S. M. (1993). Asthma and emotion: A review. *Journal of Asthma*, *30*(1), 5–21.

Listerman, T. (2010). Framing of science issues in opinion-leading news: international comparison of biotechnology issue coverage. *Public Understanding of Science*, *19*(1), 5–15. https://doi.org/10.1177/0963662505089539

Mallett, R., & Hagen-Zanker, J. (2018). Forced migration trajectories: An analysis of journey- and decision-making among Eritrean and Syrian arrivals to Europe. *Europe. Migration and Development*, *7*(3), 341–351.

Matos, S. M. A., Jesus, S. R., Saldiva, S. R. D. M., Prado, M. S., D'Innocenzo, S., Assis, A. M. O., Rodrigues, L. C., Alcantara-Neves, N. M., Cruz, Á. A., Simões, S. de M., Barreto, M. L., & SCAALA (Social Change, Asthma and Allergy in Latin America) Study Group. (2011). Overweight, asthma symptoms, atopy and pulmonary function in children of 4–12 years of age: Findings from the SCAALA cohort in Salvador, Bahia, Brazil. *Public Health Nutrition*, *14*(7), 1270–1278. https://doi.org/10.1017/S1368980011000462

McIntosh, M. J., & Morse, J. M. (2015). Situating and constructing diversity in semi-structured interviews. *Global Qualitative Nursing Research*, *2*, 2333393615597674. https://doi.org/10.1177/2333393615597674

Meyrowitz, J. (2009). We liked to watch: Television as progenitor of the surveillance society. *The Annals of the American Academy of Political and Social Science*, *625*(1), 32–48. https://doi.org/10.1177/0002716209339576

Miles, C., Arden-Close, E., Thomas, M., Bruton, A., Yardley, L., Hankins, M., & Kirby, S. E. (2017). Barriers and facilitators of effective self-management in asthma: systematic review and thematic synthesis of patient and healthcare professional views. *NPJ Primary Care Respiratory Medicine*, *27*(1), 57. https://doi.org/10.1038/s41533-017-0056-4

Noar, S. M., Harrington, N. G., Van Stee, S. K., & Aldrich, R. S. (2011). Tailored health communication to change lifestyle behaviors. *American Journal of Lifestyle Medicine, 5*(2), 112–122. https://doi.org/10.1177/1559827610387255

Ochoa-Avilés, A., Verstraeten, R., Huybregts, L., Andrade, S., Van Camp, J., Donoso, S., Ramírez, P. L., Lachat, C., Maes, L., & Kolsteren, P. (2017). A school-based intervention improved dietary intake outcomes and reduced waist circumference in adolescents: a cluster randomized controlled trial. *Nutrition Journal*, *16*(1), 79. https://doi.org/10.1186/s12937-017-0299-5

Ottoni-Wilhelm, M., Estell, D. B., & Perdue, N. H. (2014). Role-modeling and conversations about giving in the socialization of adolescent charitable giving and volunteering. *Journal of Adolescence*, *37*(1), 53–66. https://doi.org/10.1016/j.adolescence.2013.10.010

Palmieri, P. A. (2017). La técnica Delphi: Un método de consenso para la investigación en servicios de salud en Latino América. *Ágora Revista Científica*, *4*(2), 30–38. https://doi.org/10.21679/arc.v4i2.89

Riera, A., Ocasio, A., Tiyyagura, G., Krumeich, L., Ragins, K., Thomas, A., Trevino, S., & Vaca, F. E. (2015). Latino caregiver experiences with asthma health communication. *Qualitative Health Research*, *25*(1), 16–26. https://doi.org/10.1177/1049732314549474

Rimer, B. K., & Kreuter, M. W. (2006). Advancing tailored health communication: A persuasion and message effects perspective. *The Journal of Communication*, *56*(1_suppl.), S184–S201. https://doi.org/10.1111/j.1460–2466.2006.00289.x

Rodríguez, J. C. F., Muñoz, F. M., & Muñoz, A. C. (2018). Presence of negative emotions in bronchial asthma. *Revista Iberoamericana de las Ciencias de la Salud: RICS*, *7*(14), 1–20.

Stowell, E., Lyson, M. C., Saksono, H., Wurth, R. C., Jimison, H., Pavel, M., & Parker, A. G. (2018). Designing and evaluating mHealth interventions for vulnerable populations: A systematic review. In Association for Computing Machinery, *Proceedings of the 2018 CHI Conference on Human Factors in Computing Systems* (pp. 1–17).

Varela Ruiz, M., & Vives Varela, T. (2017). No es aceptación, es sembrar la semilla para entender mejor una tensión educativa. Réplica. *Investigación en educación médica*, *6*(23), 216–217. https://doi.org/10.1016/j.riem.2016.09.010

Vásquez-Guevara, D. H. (2019). *Science for all: Exploring science communication for public engagement in culturally diverse scenarios in the Americas*. (Publication No. 13812526) [Doctoral dissertation, The University of New Mexico].

Vásquez-Guevara, D. H. (2021). Ciencia para Todos: Guías para superar los Desafíos de la Comunicación Científica en Casos de Comunicación de la Salud. *Fonseca Journal of Communication*, *22*. https://doi.org/10.14201/fjc-v22-22148

Weingart, P., & Joubert, M. (2019). The conflation of motives of science communication—causes, consequences, remedies. *Journal of Science Communication*, *18*(3), Y01. https://doi.org/10.22323/2.18030401

Chapter 5

Implications for Effective Science Communication to Reach Public Engagement

Denisse Vásquez-Guevara and
Fernando Ortiz-Vizuete

We can agree that there is still a lot to advance in adopting new frameworks for science communication that motivate dialogue, interaction, and participation, with the goal of collaboration between scientists and people. In this final chapter, we provide you—whether you are a grad student, experienced scientist, or a science follower—reflections on how to think about crucial factors when incorporating science communication that aims to create meaningful engagement.

We explain how theories can be used to guide research and practice of science communication research and practical initiatives. In terms of epistemology, we explore how decolonial paradigms could be useful to create meaningful relationships between scientists and people. Our final set of reflections about key implications presents methodological guidelines reflections to conduct science communication projects and develop communication pieces.

SCIENCE COMMUNICATION GUIDELINES FOR PUBLIC ENGAGEMENT

Scientists have multiple activities, which present a variety of challenges they need to navigate every day in their research endeavors. However, now more than ever, scientists and scientific organizations need to connect better with the broader society and decision-makers, first, to inform how to prevent future pandemics, political issues, and environmental catastrophes, among other issues that affect humankind and, second, to find viable solutions through scientific data.

From the audience's perspective, we need to reflect that media and communication have evolved dramatically in the last three decades. People from multiple generational groups have gone from receiving information through traditional mass media, such as newspapers and television, to social media and digital conduits that not only allow them to learn more about the latest events but to tailor their informational sources based on their needs, interests, and demographics. This context calls for the attention of science communication research and practice, which needs to adapt its structures to be effective in informing people clearly as well as engaging them.

In this section, we present three broad guidelines to plan science communication and understand the crucial implications that are necessary to develop audience engagement.

Building Trust and Dialogic Spaces for Audiences

Before starting a science communication initiative, either research or practice, scientists and/or communication practitioners need to reflect on how to build trust with their audiences. As we have reviewed, scientists can attract an audience's attention based on their credentials and expertise (Katz, 1957). However, scientists do not frequently receive science communication training during their studies (Massarani et al., 2016), even though, public speaking training is crucial to develop science content speeches and presentations. Nevertheless, scientists must be involved and actively participate in science communication initiatives not only to demonstrate their true willingness to listen, but also to collaborate with their audiences developing practical solutions (Friedman et al., 1999).

As is evidenced in our case studies, in line with scholarship, people are open to dialogue with researchers to learn more through direct interaction as well as to have, a space to participate in topics of their own interest (Vásquez-Guevara, 2021). When researchers need to establish trust with audiences, decolonial epistemologies (as seen in chapter 2) can provide several guidelines and principles to develop procedures for research or practice that focus on opening spaces for collaboration between scientists and people.

Our studies and experiences coincide with available scholarship regarding which settings for science communication can motivate audiences to participate. In this case, audiences prefer events where they can participate actively (i.e., webinars, in-person workshops, science café) and ask their questions directly to experts. Ideal spaces to organize in-person events can be public access venues, such as parks, plazas, and coffee shops where conversations can take place as well as interpersonal interaction (Selin et al., 2017).

Science Communication Teams for Strategy and Execution

Science communication initiatives require researchers' active involvement as experts in the subject, in order to verify content accuracy and to provide professional support that can produce communication pieces. On the other hand, communication teams, depending on the format of the selected communication pieces, should include a public information officer (PIO) or science communication specialist who leads the overall strategy and copywriting processes. Moreover, PIOs' roles focus on coordinating other professionals' work, such as graphic designers to produce graphics and logos; video producers, or/and animation to produce audiovisuals that act as support pieces to represent data, situations, and specific practices to audiences; and public relations specialists to organize press events, create media kits, and provide media training to spokespeople.

Designing Frames for Science Communication Content and Messages

A key component of science communication is how we frame messages about scientific content in order to attract audiences' interest and, consequently, engage them. Again, PAR methodologies have been successful to explore collaboratively with audiences their needs and interests around specific topics (Bucchi & Trench, 2014). However, we suggest that, to complement this information, formative research is needed, as it provides demographic and psychographic data to learn more about audiences. Once we have enough knowledge about our audiences, we can also explore with them which are possible mental schemas we can use according to their age, culture, and context. As Entman (1993) presents in framing theory, schemata appeal to situations, emotions, and information that audiences can interpret and understand. However, an additional interpretation of framing developed for science communication states that building a mental schema implies combining scientific data and audience data obtained through formative research, where situations and common contexts for target audiences are combined to create a narrative that is interesting for them (White, 2013).

To develop a message and content strategy for science communication projects, we suggest the following steps based on one of our author's prior study (Vásquez-Guevara, 2021).

1. At the first meeting, present the research findings of a study to a group of your target audience, and leave enough time for dialogue and questions between the researchers or authors and the target audiences.
2. Identify the main themes of interest in the first dialogue session and invite target audiences to discuss where they can find situations or anecdotes of their lives they can relate to the research findings. For example, in our Science for All study, we asked co-researchers which moments they enjoy the most; they referred to their school recess time when they played with their friends and enjoyed lunch, and weekends when they could spend time with their families and loved ones. Consequently, the artwork context for the communication pieces had to be school settings and family homes. Moreover, we invited the program's former participants to be the faces of the communication pieces in printed media and social media.

Designing Effective Messages

Developing messages is fundamental for science communication projects and to attract the audiences' interest. Therefore, messages are the main carriers of key scientific content that will lead to audiences' understanding of a specific subject. Additionally, well-developed messages support the overall science communication strategy that can lead to audiences' engagement and the performance of expected behaviors aligned with the project's goals.

A useful tool to design science communication messages is the *Message Box* (COMPASS, 2020a), developed by science communication researchers and practitioners with diverse multidisciplinary backgrounds. They work to provide scientific engagement through public outreach solutions, training, and resources for scientists, communities, and policymakers around science communication.

The Message Box focuses on finding a scientific study's key ideas for developing a message that resonates with the target audience. The Message Box uses six criteria: *audience, problems, benefits, solutions, so what?, and issue* (COMPASS, 2020a).

Audience refers to the specific group we want to address through a message. In the case of science communication, these could be journalists who will be key in disseminating the message to broader audiences, policymakers who need to consider specific data during decision-making processes, and legislators, or community members who we need to make aware about of a specific issue that affects them. As can be seen, a project might have several audiences. However, we should run the Message Box process for each of them addressing the specific goals we have with each audience.

The *issue* criterion (in the center of the box, as seen in figure 5.1) broadly describes and identifies the main research topic. This section should summarize in one phrase the project's context and broad scope, for example: the food safety practices of Cuenca's informal food entrepreneurs in the context of the COVID-19 pandemic.

The *problem* refers to a section of the issue we are investigating through a specific discipline by conducting specific experiments or analyses. The *problem* is related to the research questions and goals, guiding the overall study that addresses an issue's critical areas. For example, the lack of training opportunities for food safety practices for entrepreneurs leads to misinformed practices of cooks and handlers. Food mishandling negatively impacts negatively food quality and, consequently, this causes food-related illnesses, such as cholera and stomach infections, which become public health problems.

The *so what* criterion refers to finding the reasons why audiences should care or become interested in a scientific study, what about a study is key for them to know, and the reasons why, as researchers, we need to talk to them. It is necessary to reflect on the audience's needs and interests and how we can frame the scientific data through their life scenarios to approach them and provide key information for our study. Here, we can appeal to how certain behaviors and practices can solve or negatively affect an existent situation. For example, food poisoning is the most common public health issue among people in Ecuador; however, people might need to become aware that it also leads to chronic illnesses such as obesity, diabetes, or heart diseases.

The criterion referring to *solutions* presents the broad solutions that are available or that the study proposes to address the problem. When solutions are presented, it is crucial to suggest actions that audiences can put into practice to contribute to the solution of the problem. For example, the Ecuadorian Research Consortium for Food Safety will offer free training sessions in which you will learn how to develop food safety practices for your business and contribute to the health of your customers and business.

Regarding *benefits,* we need to list and explain how the solutions we are proposing are useful to solve the problem. In this case, Message Box encourages people to focus on the positive scenarios that will happen if they join the implementation of the solutions, and how these efforts are meant to solve the problem.

The Message Box offers us a useful tool to explore how to develop engaging messages that not only focus on disseminating scientific knowledge but also suggest behavioral guidelines. Alternatively, science communication messages developed through the Message Box tool offer audiences useful knowledge applicable to their different life scenarios as solutions to problems that affect them directly or indirectly. Moreover, it invites audiences to reflect on how they can adhere to actions that lead to solutions to an overall issue.

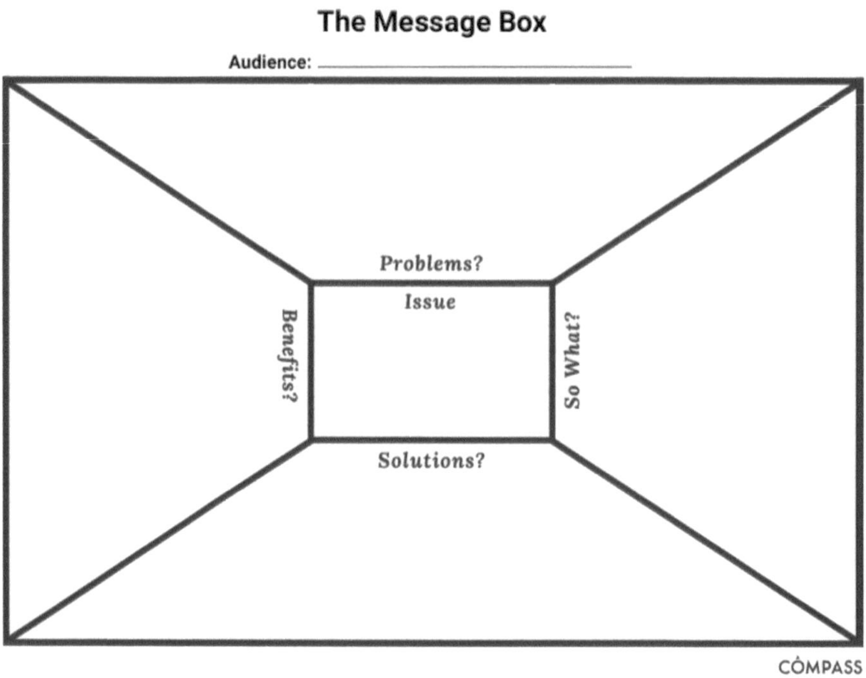

Figure 5.1. Compass message box. ***Reprinted courtesy of COMPASS (2020).***

Choosing Communication Channels and Communication Formats

As we have presented throughout this book, science communication for public engagement's priority is to motivate dialogue between the scientific community and society. However, to motivate effective dialogue with audiences, correct message formats and communication conduits must be chosen that foster audience participation with the ultimate goal of engagement.

We used the medium theory (Collins et al., 2016) as a foundation to better understand the media features, such as its formats, in order to choose communication conduits that provide specific functions to enhance the understanding of scientific content. For example, in our case studies in chapter 4, we identified specific information that needed to be presented in a format that could be revised by audiences easily and frequently. For this reason, certain data needed to be printed or presented through diagrams or figures (i.e., the caloric content of foods and their healthy portions), in a format such as a booklet or social media posts that show an infographic, that audiences could print or saved as a picture for later consultation of the audiences. However, other information can be better explained through tutorials, such as by cooking recipes videos or by creating online events (e.g., Instagram Live events), where people can ask questions to experts. For this reason, we chose to develop Instagram Reels as video tutorials that show how to prepare healthy meals step by step.

Moreover, we can suggest that social media is not the only conduit to create an impact on audiences. The disparities in access to the internet and mobile devices affect low-income and vulnerable populations. In this case, we suggest that it is necessary to combine online and mass media conduits and activities to ensure that people have access to learning and are informed about science.

Science communication is in constant and rapid evolution. For this reason, we will explore one of the latest trends in public engagement: the use of audiovisual narratives.

In the age of social media, videos, and digital communication, we will present several guidelines to use and create audiovisuals that support the dissemination and communication of science communication initiatives that are engaging and enjoyable for scientists and the broad public.

SOCIAL MEDIA AUDIOVISUAL NARRATIVES FOR SCIENCE COMMUNICATION

Social media, new technological mobile devices, and the Internet have revolutionized the way people communicate and interact every day, and in the last decade with the use of audiovisual narratives as part of social media content. Science communication is not the exception. Currently, scientists, universities, and scientific research institutions are including audiovisual narratives, and different formats to present to the broader public their work to inform and educate non-expert audiences. The multiple platforms of massive audiences use different formats that combine images, text, and video. These formats open opportunities to build and narrate complex scientific data and engage audiences. In this chapter, we will explore several guidelines to build audiovisual narratives for science communication.

THE SCARCITY OF SCIENTIFIC CONTENT IN MASS MEDIA

Traditional media outlets in radio, television, and print media offer almost none or no space for science. In the last two decades, media outlets have faced dramatic changes and economic crises that have reduced investments and operational budgets which impacted media content by reducing sections and its available variety of information (Brumfiel, 2009). Science sections and specialized reporters were removed from television, radio, magazines, and newspapers (Dunwoody, 2014) with the few exceptions of specialized media outlets such as *Nature*, *National Geographic*, or *Scientific American*.

Consequently, there are still multiple barriers that still represent challenges for science communication. These challenges affect the access and availability of scientific content for broad audiences in mass media, and the institutional barriers to developing science communication from universities and research institutions, as reviewed in the introduction to this book and chapter 2.

At the same time, with the massification of the Internet, pseudoscience content invaded social media and websites (Nisbet & Scheufele, 2009) but with less reliable facts that did not show robust sources and studies to back up the statements or arguments of the information presented to audiences.

Social Media as an Opportunity for Science Communication Pieces Production

On a positive note, social media and technology also offer multiple opportunities to present scientific data to audiences through different formats. Moreover, social media platforms, software, and technological devices offer multiple tools and formats that simplify the process to produce communication pieces and disseminating them to audiences. Before social media, producing scientific content in a video format represented a significant effort in terms of budget and resources. However, nowadays. the reduced budgets or scarce funds for developing science communication initiatives in Latin America (Massarani et al., 2012) as well as in the United States and European countries (Stilgoe et al., 2014) limit the available resources to develop science communication strategies and pieces for non-experts.

However, social media has opened many opportunities for science communication. Now platforms such as Twitter, Facebook, YouTube, and Instagram can create conduits for direct interactions among scientists and people of different age groups, cultures, and demographic groups. Available studies demonstrate that scientists that share their research outcomes and findings on Twitter and Facebook can reach broader audiences of non-experts, and also benefit from networking with citizen groups and other researchers (Pearce et al., 2015).

Moreover, video formats through platforms such as YouTube, Instagram (through its content format of Stories and Reels), and TikTok offer multiple opportunities for self-producing audiovisual content that can be created by scientists or communication practitioners themselves to invite audiences to dialogue, learn, and interact around scientific research.

Habermas (1991) proposes the idea that in the public sphere that society can be engaged in critical public debate. Moreover, public discussion requires that information and communication are accessible to all citizens in a democratic manner. In the basic notions and strict sense of the public sphere, social media provides a space to create a democratic discussion that opens space for public participation.

Since the 2000s decade, the term social media has emerged as a buzzword of marketing ideology to attract investors and creators of different platforms with the promise to connect people with different approaches such as images, micro-blogging, professional networking, and personal contacts (Fuchs, 2014). However, scientists could be reluctant to use social media because of the fear of appearing unprofessional, and the lack of training on online communicative competencies and skills to create content that communicates the scientific content (Osterrieder, 2013).

However, scientists do not need to engage in mainstream social media trends by dancing or changing their personalities to present the results of their research in an engaging manner to audiences. We believe that social media can be used as a tool to dialogue with audiences and build trust among researchers and people to build sustainable relationships and online communities around science.

Benefits of Using Social Media to Communicate Science

Available studies evidence that using social media as a tool to communicate science has positive effects on authors and engaging key audiences around specific topics. Scientists who use Twitter as a science communication can benefit their personal and professional profile, engage audiences (López-Goñi & Sánchez-Angulo, 2018), and improve the visibility and citation of their academic papers (Pearce et al., 2015). During COVID-19, the use of Facebook, Twitter, and Telegram as communication tools was highly effective to support prevention campaigns and supporting behavior change of audiences to adopt] guidelines to avoid contagion (Matta, 2020).

Instagram has also been used as a tool for archaeologists to educate the public about pervasive conspiracy theories, guidance to students, and elucidate the scientific labor of archeologists in collecting artifacts (Caspari, 2022).

Social media can be used effectively to engage audiences and create communities around topics of interest with digital literacy skills and guidelines to produce scientific content through mobile devices and apps. However, at the same time as learning social media skills, it is crucial to understand how to build narratives to explain science clearly to deliver engaging and effective messages to audiences.

Developing Engaging Messages for Science Communication

As presented in prior chapters, current science communication research and practical initiatives claim for involving audiences actively in dialogue and mutual collaboration. Social media can be a useful tool to create strategies to create a dialogue with society to bring science closer to people (Romero-Luis et al., 2020).

Social media offers different formats such as images, video, multimedia, and microblogs that are attractive to audiences of different age groups, cultures, ideologies, and personal interests due to the varied content these platforms have to offer (López-López et al., 2019). These formats make social media a powerful tool to inform and entertain audiences, which also can be used successfully to communicate science.

Available studies evidence that audiences around the globe prefer social media to learn more about science and technological advancements. The Spanish Foundation for Science and Technology (FECYT) revealed in their 2018 report, that Spaniards prefer social media (75.7%) and videos (61.9%) to learn more about the latest results in the scientific research (FECYT, 2018).

A survey study developed in the United States, United Kingdom, Australia, New Zealand, and Canada about how scientists use social media. The study revealed that scientists preferred Facebook (88%) and Twitter (30%) to share information about their projects with broader audiences (Collins et al., 2016). Regarding content, this study revealed that scientists use Facebook for sharing their experiences in the research field through posting pictures and text captions. Moreover, scientists used Facebook to identify trends that are attractive to broader audiences to use these as ideas to communicate science, such as memes, connecting other researchers, and making corrections to misinformation or fake news around the science of their specific field.

Regarding Twitter, scientists use this microblogging platform to share specific messages about their research, such as statistics or sharing their research publications. Moreover, scientists recognized that they use Twitter to promote their work and network with other scientists through hashtags (Collins et al., 2016).

As we can see audiences and scientists use social media for different purposes. Audiences get informed and entertained through different platforms, and scientists use social media to share their work. However, developing engaging content for science communication through social media can be more feasible through several guidelines.

Guidelines for Developing Scientific Content for Social Media

In this section, we will present broader guidelines for planning science communication content audiovisual narratives.

First, narratives are stories about events or things that occur at a specific time and unfold through different causes and consequences with the participation of different people (Mandelbaum, 2003). Science narratives have been used under this assumption to educate people, support behavior change, and raise awareness about social issues (Papa et al., 2000). Some of the formats that science narratives had used in mass media are soap operas, comic books, music, and TV shows. Currently, social media has used science narratives to develop content strategies by combining contextual information that is relatable to audiences through their needs and interests with scientific data (Dahlstrom, 2014).

Some guidelines to choose spokespeople, formats, and messaging strategies to deliver engaging messages are the following.

Spokespeople. To present scientific content to audiences, choosing a trustable spokesperson is crucial for attracting the audience's attention. Science communication studies reveal that audiences are more open to listening and accepting recommendations from scientists or credentialed experts, especially if the suggested behaviors or actions imply health, and safety practices (Vásquez-Guevara, 2019; Vásquez-Guevara, 2021).

Consequently, if scientists can be present as the main carriers of the message, they must consider not overusing technical terms, or explaining their meaning to audiences. Other valuable spokespeople are people who represent the audience, who can appear in social media posts or any format by sharing their experiences or be the companion of the expert. The scientists can deliver the scientific content, and audiences can demonstrate positive practices in their environments (i.e., take medication, recycle plastics, tell their experiences of having a medical condition or share their stories of overcoming hurdles).

However, it is not recommendable to use impersonal messaging strategies, such as using voice-overs that narrate a practice without an identifiable individual or expert or using image banks to illustrate similar suggested practices. These strategies tend lower audiences' engagement and interest as they cannot see themselves identified in the message. Instead of using banks for video, multimedia, or images as resources to create a message, researchers can invite the participants of their studies to participate in science communication pieces with them as spokespeople. Available research reveals that role modeling strategies (Fogg-Rogers et al., 2017) (i.e., using success stories with people that followed specific positive behaviors) and peer education (Peykari et al., 2011) (i.e., people who are patients or use their testimonials to share their stories) in science and health communication are effective strategies to engage audiences and persuade them for behavior change and vicarious modeling, which promotes learning through observation of others practices and stories.

Messaging Strategies. As mentioned in chapter 1 and chapter 4, framing theory explains how people make sense of the information they receive through mental schemes or frames they have in their minds to make sense of information. To translate this into practice, communication practitioners and scientists can identify situations, and experiences that are frequent or common to their target audiences to find a suitable context to use as a frame to present scientific content. Formative research that presents sociographic, demographic, and psychographic data about the target audiences will be useful for this purpose.

To develop an efficient messaging strategy, after identifying possible contexts or situations to frame the scientific content, it is crucial to identify the best channels to deliver the messages and to present the data clearly for the audience's understanding (Lundgren et al., 2022). Moreover, in the case of choosing suitable social media, appropriate communication platforms are crucial for science communication messaging strategies' success. Video formats are useful to present, explain and educate people, but image-based formats can be used as a complement to remind audiences of key information. An additional consideration to choose social media conduits is to have in mind the platforms that are mostly used by the target audiences.

Once the message framing and the appropriate communication conduits are chosen it is crucial to design the messages according to each social media platform. Other techniques for message design, however, are storytelling and participatory methodology approaches. Storytelling has proven to be a highly effective technique to present scientific content. Recent studies evidence the potential of storytelling to create emotional connections with audiences to make them care about science, beyond presenting facts, and evidence (Joubert et al., 2019). Storytelling uses the *narrative arc* as a structure to develop a story with a beginning, a middle with some conflict or issue known as a plot, and an end that shows results or solutions. To attract the audience's attention twists and turns are used to characterize a storyline that is engaging (Storr, 2019).

To develop effective storytelling strategies for science communication strategies, practitioners must focus on fostering audience collaboration to enhance the effectiveness of the overall messaging strategies. In chapter 3, this book presents several methodological considerations and case studies that illustrate different guidelines.

Social Media Posts Formats Recommendations. Social media platforms offer different formats that combine video, music, images, text, and multimedia. Platforms such as Instagram, YouTube, Tik Tok combine video and tools for video editing to create content. Moreover, through different types of posts such as stories, and reels, which are audiovisual narratives. Other platforms such as Facebook, as well as Instagram, can use audiovisuals, images, and text.

Some studies that evaluate science communication content effectiveness in social media for public engagement recommend the following guidelines to develop engaging messages.

Writing Text for Posts. A social media post can use text captions to accompany images and videos. For developing a text caption, it is recommended to be as succinct as possible and avoid long texts. It is recommended to start with a hook that attracts the audience's attention such as a question, a quote, or a situation, to introduce the topic briefly (Cann et al., 2011). Audiovisual social media such as TikTok, Instagram, and YouTube, use videos and/or images, these features must be used to develop clear and focused explanations with a video or image that presents clear information, and recommendations to audiences.

Engaging social media posts use hashtags and emojis that enhance text content, as it can be useful to make specific data more salient to social media users.

Hashtags and URLs. Facebook and Twitter posts that use hashtags as part of the text posts improve audience interaction. Moreover, studies evaluating engagement of science communication posts evidenced that if the posts include URL links to more information have higher interaction than posts that only include text (Lundgren et al., 2022). Hashtags allow social media users to develop specific searches about topics of interest in each social media platform. For this reason, using hashtags promotes better audience engagement and interaction, because it connects audiences with specific information interests.

Content Considerations. To present research data, it is crucial to develop social media posts with specific topics and purposes. For this reason, planning content according to the communication goals is crucial. Presenting an issue can require several messages to explain clearly its implications. A useful resource and post format are infographics which can be useful to present information clearly by using color, text, and graphics to explain an issue (Huang et al., 2019).

Another consideration to develop social media content for science is to keep in mind the focus of the expected outcomes of the science communication strategy. For example, if a research project aims to raise awareness about climate change and mitigation strategies, the social media strategy must develop content that explains the scientific facts and evidence and its impact on the environment, simultaneously it is crucial to present recommendations for audiences about how they can take action in contributing to the mitigation of the problem.

Consequently, science content must not only inform or raise awareness, but must present feasible actions, or practices that can be developed by audiences to address the issues, prevent further complications and find solutions. Presenting suggested behaviors can be interpreted as a *call to action*, that in strategic communication motivates audiences to pursue specific behaviors (FitzPatrick Sifford & Cohen-Aponte, 2019). For science communication, social media content frequently uses tutorials or DIY content that explain to audiences step-by-step how to do different activities, such as cooking a healthy meal, recycling, and health practices such as the correct use of face masks or hand washing to prevent COVID-19.

New media is incorporating different communication narratives that combine audiovisuals and technology to communicate, entertain, and educate audiences. The next section will explore how audiovisual narratives can be used to develop scientific content for science communication.

Audiovisual Narratives for Science Communication in Social Media

There are different types of audiovisual narratives used for science communication, some of these are personification, presentation, expert videos, animation, and narrative explanatory narratives.

Personification

As a narrative, personification borrows some values and practices from journalism and specifically from the news reporting (Dahlstrom, 2014). Through personification, individuals develop reports and explain scientific content through different resources such as presenting examples, and scientific evidence from studies to contextualize a specific topic. YouTubers or TikTok content creators use the narrative of personification to narrate scientific facts to broader audiences.

Social media and the internet have transformed the way science communication can be produced and presented to the public. Audiences are no longer passive recipients of scientific information, but rather active participants in the process of dialoguing with scientists directly through social media platforms (Boy et al., 2020).

In Latin America, Aldo Batra through his YouTube channel Plato's Robot develops videos around topics of public interest to explain issues such as COVID-19, space science, and climate change. Batra is a professional science communication practitioner from Peru, based in New Zealand, and has over 2.3 million followers. Bartra uses himself as a narrator, to present the topic and center the audience's attention. This strategy is used by media, and science communication practitioners accompanied also by storytelling tools to engage non-expert audiences (Badenschier & Wormer, 2012).

Moreover, Batra motivates his audiences to suggest him topics of interest to develop his videos and dialogues actively to his audiences through the comments that YouTube users leave him in his videos. Additionally, Batra dedicates time in most of the videos to respond to audiences about common questions and comments. Engaging in dialogue with audiences through social media is a key factor in building audience engagement and online communities. Moreover, this phenomenon known as parasocial interactions explains how audiences build mediated relationships with content creators that give an illusion of face-to-face relationships that are very influential on audiences' behavior (Papa et al., 2000).

Scientists and scientific organizations can foster and motivate audiences' engagement by responding, commenting, and interacting actively with audiences. If scientists develop sustained relationships with their audiences, they have the opportunity to build communities that are interested in scientific research and specific topics that affect society directly.

Presentation

Presentation videos are classical lectures, frequently these present the scientists and experts in a medium close-up looking at the camera to present a topic, respond to frequent questions, or clarify misinformation (Boy et al., 2020).

Expert

Narratives with experts focus attention on a person that is an attention hook to present a topic (Boy et al., 2020); these can use animation and graphics as support resources to explain certain information. Frequently, these videos are used as PR strategies for universities or research institutions.

Animation

Videos with animation using computer-generated images and simulations to illustrate a process, a problem, or scientific procedures accompanied by voiceover narrations or captions to support audiences' understanding.

Narrative Explanatory

Narrative explanatory videos are led by a specific question that would be answered throughout the video through different resources such as expert interviews, animations, and other resources from the three types detailed before (Boy et al., 2020). Narrative exploratory videos use logical reasoning to provide arguments to explain why the specific phenomenon happens and exists, their causes and consequences, and different alternatives to solve or address the issue.

Developing Engaging Audiovisual Narratives for Science Communication

Available studies show that there are specific elements in social media videos that enhance audience engagement and interaction. Regarding science communication videos, some studies have explored how marketing communications guidelines can borrow elements used in commercials to develop storytelling pieces around scientific data. Finkler and Leon-Anguiano (2019) adopted the model of SUCCESS developed by Heath and Heath (2007) and Shimp and Andrews (2013) to develop engaging advertising and applied these principles to develop science communication videos about whale watching research. Results of the study evidenced improvements in audiences' attention and engagement.

The SUCCESS model elements adapted to science communication audiovisual narratives are the following (Finkler & León-Anguiano, 2019).

Simplification

Effective science communication keeps content simple and clear. Simplification's core idea is that a study must be used as a hook to draw people's attention to support the audience's learning and remember key information. Here is not necessary to give detailed information about a study, but rather center on its main goals and implications. For example, on COVID-19 transmission, the main purpose will be to show audiences how the virus can be passed from one person to other, and how this can be avoided.

Unexpectedness

Science content must focus on offering information that is unknown to audiences. Consequently, science content must communicate new information that invites the audience to discover and reveal facts they do not know. For this purpose, audiovisual narratives for science must deviate or overcome audiences' prior expectations. A useful resource could be to present a question and answer it with information that raises the audience's awareness and attention. For example, show a full-screen caption with the question *How can COVID-19 prevention help you?* The next frames could show which related illnesses could be avoided by following guidelines or avoiding contagion.

Concreteness

While developing texts and content science communication must avoid using intangible concepts that are not commonly known by audiences. Instead, it is recommended to use tangible practices that are related to concrete actions, qualities, and characteristics that can be feasible actions (i.e., washing hands or keeping surfaces clean). For scientists it is particularly difficult to translate abstract concepts into concrete examples. For this reason, the support of science communication practitioners to develop animations, graphics, or examples that support audiences' understanding and learning processes. A useful resource is to use text captions and images of key concepts to support audiences to remind these concepts, their meaning, and derived implications.

Credibility

Science content must focus to make data and information believable to audiences. For this purpose, the narrative should build arguments with valid data and facts that are real for to audiences accept these as facts. A key factor for credibility is the likeability of the content and spokespeople that appear in the science narratives. For this reason, experts that appear in the narrative or present science content must be credentialed professionals, but also people that appear must represent audiences to increase narrative persuasion.

This element supports the prior information presented as criteria to select spokespeople for the science content presented before in this chapter.

Emotions

Credibility also has a strong relation to the authenticity and trustworthiness presented in the narrative by showing real human reactions and emotions that can reflect audiences' reactions that are natural and honest. For example, to communicate COVID-19 prevention, it might be necessary to appeal to emotions such as love to friends and family and significant others through showing images of people taking care of others and themselves by following prevention guidelines. Moreover, showing new ways for social interactions such as meeting in outdoor spaces and keeping a social distance to interact with others can be portrayed as new ideas to stay in touch with beloved ones to prevent COVID-19.

Positive emotions such as hope and awe benefit long-term audiences' receptiveness to science content. Specifically, to motivate behavior change, these positive emotions must be combined with step-by-step practices and actions as solutions that audiences can put into practice to be part of broader efforts to drive change (Garland et al., 2010). Instead, negative emotions such as sadness or fear might be necessary resources for risk and health communication topics. However, these are frequently unlikely to motivate the virality of videos (Berger & Milkman, 2012).

Science

The heart of science communication content is to present scientific data creatively to engage audiences. For achieving engagement as a goal, a successful technique is developing storytelling hooks by linking scientific data with situations from people's everyday lives. For example, on COVID-19 preventive communication, audiovisual narratives can present medical experts, families, educators, and health staff doing their work and supporting each other by following specific activities such as nurses taking care of ill people, children that correctly use face masks, and doctors administrating vaccines to people of different ages.

Storytelling for Science

Science stories can follow specific guidelines to present science stories. Storytelling for science communication narratives can be informed through the ABT (*and, but, therefore*) framework (Olson, 2009). *And* refers to the linked events or phenomena to a current issue and situation through a cause-and-effect logic. *But* presents the available evidence, data, facts, and events around the main topic to interpret the reality. Finally, *therefore*, suggests what can be done to address the issue through scientific data into feasible actions that can be executed or put into practice by audiences to be part of the solution.

For example, the ABT model adapted to the COVID-19 prevention campaign: COVID-19 has infected millions of people around the world *and* still affects and spreads from person to person, *but* research shows, that reinfection causes different long-term consequences on health that can make people more exposed to develop chronic illnesses, *therefore,* to prevent unknown consequences on health sustaining health prevention guidelines is crucial to mitigate the long-term impact of COVID-19.

Producing Content through Social Media Features and Tools

Social media platforms offer different tools to edit videos, images, and audio through smartphone apps. Instagram, YouTube, and TikTok offer video edition tools to create content that includes music trends, gifs, stickers, image filters, and more. These features make content production more accessible to scientists and science communication practitioners.

Moreover, apps to edit videos and images, such as In Shot, and to create artwork designs and animation, such as Canva and Unfold, can be useful tools to create audiovisual narratives for social media posts. These tools make content creation more accessible for science communication, in terms of budgets and funding. Scientists and science communication practitioners can create content for social media such as stories, reels, and short videos to reach audiences with different information and recommendations for broader audiences.

However, to create more elaborated communication pieces that are created with good quality it is necessary to include videographers, animation designers, and design professionals as part of the science communication team.

BIBLIOGRAPHY

Badenschier, F., & Wormer, H. (2012). Issue selection in science journalism: Towards a special theory of news values for science news? In S. Rödder M. Franzen, & P. Weingart (Eds.), *The sciences' media connection: Public communication and Iits repercussions. Sociology of the sciences yearbook, vol. 28* (pp. 59–85). Springer Netherlands. https://doi.org/10.1007/978-94-007-2085-5_4

Berger, J., & Milkman, K. L. (2012). What makes online content viral? *Journal of Marketing Research*, *49*(2), 192–205. https://doi.org/10.1509/jmr.10.0353

Boy, B., Bucher, H.-J., & Christ, K. (2020). Audiovisual science communication on TV and YouTube. How recipients understand and evaluate science videos. *Frontiers in Communication*, *5*. https://doi.org/10.3389/fcomm.2020.608620

Brumfiel, G. (2009). Science journalism: Supplanting the old media? *Nature*, *458*(7236), 274–277. https://doi.org/10.1038/458274a

Bucchi, M., & Trench, B. (Eds.). (2014). *Handbook of public communication of science and technology*. Routledge.

Cann, A., Dimitriou, K., & Hooley, T. (2011). *Social media: A guide for researchers*. Research Information Network.

Caspari, G. (2022). Instagram as a tool for archaeological science communication. *Digital Applications in Archaeology and Cultural Heritage*, *24*, e00219. https://doi.org/10.1016/j.daach.2022.e00219

Collins, K., Shiffman, D., & Rock, J. (2016). How are scientists using social media in the workplace? *PloS One*, *11*(10), e0162680. https://doi.org/10.1371/journal.pone.0162680

COMPASS. (2020a). *The message box workbook*. Compass. chrome-extension://efaidnbmnnnibpcajpcglclefindmkaj/https://www.compassscicomm.org/wp-content/uploads/2020/05/The-Message-Box-Workbook.pdf

COMPASS. (2020b, April 29). *The message box*. COMPASS. https://www.compassscicomm.org/leadership-development/the-message-box/

Dahlstrom, M. F. (2014). Using narratives and storytelling to communicate science with nonexpert audiences. *Proceedings of the National Academy of Sciences of the*

United States of America, 111 (4_suppl.), 13614–13620. https://doi.org/10.1073/pnas.1320645111

Dunwoody, S. (2014). Science journalism: Prospects in the digital age. In M. Bucchi & B. Trench (Eds.), *Handbook of public communication of science and technology* (pp. 43–55). Routledge.

Entman, R. (1993). Framing: Toward clarification of a fractured paradigm. *Journal of Communication, 43*(4), 51–58.

FECYT. (2018). *Percepción social de la ciencia y la tecnología en españa. Ministerio de ciencia, innovación y universidades*. FECYT. https://icono.fecyt.es/sites/default/files/filepublicaciones/18/epscyt2018_informe_0.pdf

Finkler, W., & León-Anguiano, B. (2019). The power of storytelling and video: A visual rhetoric for science communication. *Journal of Science Communication, 18*(5). https://dadun.unav.edu/handle/10171/59881

FitzPatrick Sifford, E., & Cohen-Aponte, A. (2019). A call to action. *Art Journal, 78*(4), 118–122. https://doi.org/10.1080/00043249.2019.1684113

Fogg-Rogers, L., Sardo, M., & Boushel, C. (2017). "Robots vs animals": Establishing a culture of public engagement and female role modeling in engineering higher education. *Science Communication, 39*(2), 195–220. https://doi.org/10.1177/1075547017696169

Friedman, S. M., Dunwoody, S., & Rogers, C. L. (1999). *Communicating uncertainty: Media coverage of new and controversial science*. Routledge.

Fuchs, C. (2014). Social media and the public sphere. *TripleC: Communication, Capitalism & Critique. Open Access Journal for a Global Sustainable Information Society, 12*(1), 57–101.

Garland, E. L., Fredrickson, B., Kring, A. M., Johnson, D. P., Meyer, P. S., & Penn, D. L. (2010). Upward spirals of positive emotions counter downward spirals of negativity: Insights from the broaden-and-build theory and affective neuroscience on the treatment of emotion dysfunctions and deficits in psychopathology. *Clinical Psychology Review, 30*(7), 849–864. https://doi.org/10.1016/j.cpr.2010.03.002

Habermas, J. (1991). *The structural transformation of the public sphere: An inquiry into a category of bourgeois society*. MIT Press. https://play.google.com/store/books/details?id=e799caakIWoC

Heath, C., & Heath, D. (2007). *Made to stick: Why some ideas survive and others die*. Random House.

Huang, G., Li, K., & Li, H. (2019). Show, not tell: The contingency role of infographics versus text in the differential effects of message strategies on optimistic bias. *Science Communication, 41*(6), 732–760. https://doi.org/10.1177/1075547019888659

Joubert, M., Davis, L., & Metcalfe, J. (2019). Storytelling: The soul of science communication. *Journal of Science Communication, 18*(5), E. https://doi.org/10.22323/2.18050501

Katz, E. (1957). The two-step flow of communication: An up-to-date report on an hypothesis. *Public Opinion Quarterly, 21*(1), 61. https://doi.org/10.1086/266687

López-Goñi, I., & Sánchez-Angulo, M. (2018). Social networks as a tool for science communication and public engagement: Focus on Twitter. *FEMS Microbiology Letters*, *365*(2). https://doi.org/10.1093/femsle/fnx246

López-López, E., Tobón, S., & Juárez-Hernández, L. G. (2019). Escala para Evaluar Artículos Científicos en Ciencias Sociales y Humanas-EACSH. *REICE Revista Iberoamericana Sobre Calidad Eficacia y Cambio En Educación*, *17*(4), 111. https://doi.org/10.15366/reice2019.17.4.006

Lundgren, L., Crippen, K. J., & Bex, R. T., II. (2022). Social media interaction as informal science learning: A comparison of message design in two niches. *Research in Science Education*, *52*(1), 1–20. https://doi.org/10.1007/s11165-019-09911-y

Mandelbaum, J. (2003). How to "do things" with narrative: A communication perspective on narrative skill. In J. Greene & B. Burleson (Eds.), *Handbook of communication and social interaction skills* (pp. 613–652). Lawrence Erlbaum Associates.

Massarani, L., Amorim, L., & de Oca, A. M. (2012). Periodismo científico: Reflexiones sobre la práctica en América Latina. *Chasqui: Revista Latinoamericana de Comunicación*, *120*, 73–77. https://dialnet.unirioja.es/servlet/articulo?codigo=5791303

Massarani, L., Reynoso-Haynes, E., Murrielo, S., & Castillo, A. (2016). *Posgrado en Comunicación de la Ciencia en América Latina: un mapa y algunas reflexiones* [Science communication postgraduate studies in Latin America: A map and some food for thought]. *Journal of Science Communication*, *15*(5), A03. https://doi.org/10.22323/2.15050203

Matta, G. (2020). Science communication as a preventative tool in the COVID-19 pandemic. *Humanities and Social Sciences Communications*, *7*(1), 1–14. https://doi.org/10.1057/s41599-020-00645-1

Nisbet, M. C., & Scheufele, D. A. (2009). What's next for science communication? Promising directions and lingering distractions. *American Journal of Botany*, *96*(10), 1767–1778. https://doi.org/10.3732/ajb.0900041

Olson, R. (2009). *Don't be such a scientist: Talking substance in an age of style*. Island Press.

Osterrieder, A. (2013). The value and use of social media as communication tool in the plant sciences. *Plant Methods*, *9*(1), 26. https://doi.org/10.1186/1746-4811-9-26

Papa, M. J., Singhal, A., Law, S., Pant, S., Sood, S., Rogers, E. M., & Shefner-Rogers, C. L. (2000). Entertainment-education and social change: An analysis of parasocial interaction, social learning, collective efficacy, and paradoxical communication. *The Journal of Communication*, *50*(4), 31–55. https://doi.org/10.1111/j.1460–2466.2000.tb02862.x

Pearce, W., Brown, B., Nerlich, B., & Koteyko, N. (2015). Communicating climate change: conduits, content, and consensus. *Wiley Interdisciplinary Reviews. Climate Change*, *6*(6), 613–626. https://doi.org/10.1002/wcc.366

Peykari, N., Tehrani, F. R., Malekafzali, H., Hashemi, Z., & Djalalinia, S. (2011). An experience of peer education model among medical science university students in

Iran. *Iranian Journal of Public Health, 40*(1), 57–62. https://www.ncbi.nlm.nih.gov/pubmed/23113055
Romero-Luis, J., Carbonell-Alcocer, A., & Barrio, M. (2020). El video artículo multimedia interactivo, un formato innovador para la comunicación científica. *Paradigmas de La Narrativa Audiovisualos. ASRI, 18.*
Selin, C., Rawlings, K. C., de Ridder-Vignone, K., Sadowski, J., Altamirano Allende, C., Gano, G., Davies, S. R., & Guston, D. H. (2017). Experiments in engagement: Designing public engagement with science and technology for capacity building. *Public Understanding of Science, 26*(6), 634–649. https://doi.org/10.1177/0963662515620970
Shimp, T. A., & Andrews, J. C. (2013). *Advertising, promotion and other aspects of integrated marketing communications*. Thomson South-Western.
Stilgoe, J., Lock, S. J., & Wilsdon, J. (2014). Why should we promote public engagement with science? *Public Understanding of Science, 23*(1), 4–15. https://doi.org/10.1177/0963662513518154
Storr, W. (2019). *The science of storytelling*. William Collins.
Vásquez-Guevara, D. H. (2019). *Science for all: Exploring science communication for public engagement in culturally diverse scenarios in the Americas* (Publication No. 13812526) [Doctoral dissertation, The University of New Mexico].
Vásquez-Guevara, D. H. (2021). Ciencia para Todos: Guías para superar los Desafíos de la Comunicación Científica en Casos de Comunicación de la Salud. *Fonseca Journal of Communication, 22*. https://doi.org/10.14201/fjc-v22-22148
White, J. (2013). Translating science, health and technology: Reporters as knowledge transfer intermediaries. *Online Journal of Communication and Media Technologies, 3*(2), 227.

Index

Page references for figures and tables are italicized.

About the Contributors

Angelica Ochoa-Aviles, PhD is main professor, College of Chemical Sciences, Universidad de Cuenca; senior scientist, Biosciences Department, Universidad de Cuenca; associate editor of the Public Health Nutrition Journal; and Secretary of the Latin American Society of Nutrition (SLAN). Dr. Ochoa's research interests are public health, prevention interventions, nutrition, and physical activity.

Fernando Ortiz-Vizuete, MST, PhD (C) is main professor, School of Journalism Universidad de Cuenca; and Dean of the College of Philosophy, Letters, and Educational Sciences. His research interests are communication and media studies, journalism, communication and technology, and media and culture.

Denisse Vásquez-Guevara, PhD is assistant professor in the Communication Department at Cal Poly Pomona; before, Dr. Vásquez-Guevara had been an associate professor in the School of Communication of the University of Cuenca in Ecuador, and Chairwoman of the Honores Vanguardia Program. She holds a PhD in Latin American & Iberian Studies from the University of New Mexico. As a science communication researcher and scholar, she has collaborated with over ten research grants and projects in public health, biosciences, urban studies, and assessment studies for preventive health campaigns of international NGOs in Latin America.

This book constitutes her second main author and editor contribution to science communication for public engagement. Dr. Vásquez-Guevara is an active researcher with universities based in the United States and Europe.

David Weiss, PhD is associate professor in the Department of Communication & Journalism and Director of Liberal Arts & Integrative Studies at the University of New Mexico. His research interests include media studies, political communication, and science and health communication.